Holger Rapp

Reconstruction of Specular Reflective Surfaces using Auto-Calibrating Deflectometry

Schriftenreihe
Institut für Mess- und Regelungstechnik,
Karlsruher Institut für Technologie (KIT)

Band 023

Eine Übersicht über alle bisher in dieser Schriftenreihe erschienenen Bände
finden Sie am Ende des Buchs.

Reconstruction of Specular Reflective Surfaces using Auto-Calibrating Deflectometry

by
Holger Rapp

Dissertation, Karlsruher Institut für Technologie
Fakultät für Maschinenbau
Tag der mündlichen Prüfung: 20. August 2012
Referenten: Prof. Dr.-Ing. C. Stiller, Prof. Dr. rer. nat. B. Jähne

Impressum

Karlsruher Institut für Technologie (KIT)
KIT Scientific Publishing
Straße am Forum 2
D-76131 Karlsruhe
www.ksp.kit.edu

KIT – Universität des Landes Baden-Württemberg und nationales
Forschungszentrum in der Helmholtz-Gemeinschaft

KIT Scientific Publishing 2013
Print on Demand

ISSN 1613-4214
ISBN 978-3-86644-966-4

Reconstruction of Specular Reflective Surfaces using Auto-Calibrating Deflectometry

Zur Erlangung des akademischen Grades

Doktor der Ingenieurwissenschaften

der Fakultät für Maschinenbau
Karlsruher Institut für Technologie (KIT)

genehmigte

Dissertation

von

DIPL.-PHYS. HOLGER H. RAPP

aus Düsseldorf

Hauptreferent: Prof. Dr.-Ing. C. Stiller
Korreferent: Prof. Dr. rer. nat. B. Jähne
Tag der mündlichen Prüfung: 20. August 2012

Preface

This thesis was written while I was employed at the Institut für Mess- und Regelungstechnik at the Karlsruhe Institute für Technologie (KIT) in Karlsruhe. It was supervised by Prof. Dr.-Ing. Christoph Stiller to whom I express my gratitude for offering me the opportunity to research deflectometry and to entrust me with this topic. I am especially grateful for the freedom he granted me for my work and for his support in my personal growth and side projects.

I am also very happy that Prof. Dr. rer. nat. Bernd Jähne has agreed to be the second examiner. The discussions with him and his team about topics of deflectometry and otherwise have helped me tremendously. I owe him my gratitude.

Of course I also want to say thanks to all the people who helped me finishing this work: My colleagues – especially my scrum partners Miriam Schönbein, Philip Lenz, and Henning Lategahn – for hours of discussion, fun, moral support and coffee; the ladies from the secretariat – namely Sieglinde Klimesch, Erna Nagler, and Silke Rittershofer – which helped me solving the really difficult tasks: German bureaucracy; Werner Paal for his assistance in all things IT and the men from our workshop who made my experimental setup not only functional, but also pretty.

It is also important to me to acknowledge the proofreaders of this work which were invaluable and made this thesis easier to read and understand. All remaining errors are of course my own.

I thank my parents and my brother for their never ending support. And I thank my companion Hanna Podewski. She has always been the sunshine of my life since I got to know her.

Karlsruhe in Mai 2012   Holger Rapp

Abstract

The three-dimensional reconstruction of highly specular reflective parts with optical measurement methods is challenging and intriguing. It is challenging because such parts can be hard to distinguish from their surroundings with an optical sensor since they do not provide texture of their own but only reflect the objects around them. The endeavour is intriguing though because there is a well established method for qualitative visual inspection for specular surfaces which offers good prospects for development into a precise quantitative reconstruction method: namely deflectometry.

This thesis discusses deflectometry as a reconstruction method for highly reflecting surfaces. It focuses on deflectometry alone and does not use other reconstruction techniques to supplement with additional data. It explains the measurement process and principle and provides a crash course into an efficient mathematical representation of the principles involved. Using this, it reformulates existing three-dimensional reconstructing methods, expands upon them and develops new ones. Building on these novel techniques, an auto-calibration is introduced that is able to refine a rough extrinsic calibration.

All methods are experimentally verified and compared with each other using simulations and experiments.

Keywords: Deflectometry, reflection methods, specular surfaces, automated visual inspection, structured lighting, 3D reconstruction, image processing, optimization, calibration

Kurzfassung

Die dreidimensionale Rekonstruktion von stark spiegelnden Objekten mit optischen Messmethoden is herausfordernd und faszinierend. Sie ist herausfordernd, da Objekte dieser Art mit optischen Sensoren nur schwer von ihrer Umgebung unterscheidbar sein können, da sie keine eigene Textur besitzen, sondern nur ihre Umgebung widerspiegeln. Sie ist aber auch faszinierend, da es bereits eine gut etablierte Methode für die qualitative Sichtprüfung reflektierender Oberflächen gibt, die gute Aussichten hat, in eine hochgenaue Rekonstruktionsmethode entwickelbar zu sein: die Deflektometrie.

Diese Dissertation diskutiert die Deflektometrie als Rekonstruktionsverfahren für stark reflektierende Oberflächen. Sie konzentriert sich ausschließlich auf die Deflektometrie und nutzt kein anderes Verfahren zur Unterstützung. Sie erklärt den Messprozess und das Messprinzip und bietet einen Crash Kurs für die effiziente mathematische Repräsentation der zugrundeliegenden Prinzipien. Unter deren Nutzung werden existierende Rekonstruktionsverfahren neu formuliert, erweitert und neue entwickelt. Mit diesen neuen Techniken wird eine Selbstkalibrierung eingeführt, die es ermöglicht grobe extrinsische Kalibrierungen zu verbessern.

Alle Methoden werden experimentell überprüft und miteinander verglichen. Hierzu werden Simulationen und reale Messdaten verwendet.

Schlagworte: Deflektometrie, Reflexionsverfahren, spiegelnde Oberflächen, automatische Sichtprüfung, strukturierte Beleuchtung, 3D-Rekonstruktion, Bildverarbeitung, Optimierung, Kalibrierung

Contents

Contents

Symbols

Abbreviations

Notation

Scalar values	non bold	$a, b, c, \ldots$
Dual scalars	non bold, with breve	$\breve{a}, \breve{b}, \breve{c} \ldots$
Vectors	bold	$\mathbf{a}, \mathbf{b}, \mathbf{c}, \ldots$
Dual vectors	bold, with breve	$\breve{\mathbf{a}}, \breve{\mathbf{b}}, \breve{\mathbf{c}}, \ldots$
Quaternions	bold, fraktur	$\mathfrak{a}, \mathfrak{b}, \mathfrak{c}, \ldots$
Dual quaternions	bold, fraktur, with breve	$\breve{\mathfrak{a}}, \breve{\mathfrak{b}}, \breve{\mathfrak{c}}, \ldots$
Matrices	bold, uppercase	$\mathbf{A}, \mathbf{B}, \mathbf{C}, \ldots$
Constants	text font, non bold	$a, b, c, \ldots$

Symbols

ϵ	Nilpotent base element for the dual constructions, $\epsilon^2 = 0$
$\mathbb{R}$	The field of real values
$\mathbb{R}^+$	Positive real values. $\mathbb{R}^+ := \{r \in \mathbb{R} \mid r \geqslant 0\}$
Δ	Abelian ring of dual numbers
$\mathbf{a} \cdot \mathbf{b}$	Scalar product between $\mathbf{a}$ and $\mathbf{b}$
$.^\mathsf{T}$	Transposition of a vector or matrix
$\mathrm{Sc}\,(.)$	Scalar part of a quaternion
$\mathrm{Vec}\,(.)$	Vector part of a quaternion
$\mathrm{Real}\,(.)$	Real part of a dual construction
$\mathrm{Dual}\,(.)$	Dual part of a dual construction
$\mathcal{H}$	The space of quaternions
$\mathbf{a} \times \mathbf{b}$	Cross product between $\mathbf{a}$ and $\mathbf{b}$
$\mapsto$	Mapping
$[\mathbf{a}]_\times$	Matrix, such that $[\mathbf{a}]_\times \mathbf{b} = \mathbf{a} \times \mathbf{b}$ for all $\mathbf{b}$.
$:=$	Definition
$\bar{.}$	Conjugate
$\|.\|$	Norm
$\bar{\bar{.}}$	Complete conjugate
$\det\,(.)$	Determinant

$\mathbf{u}_c$	Pixel index of camera
$\mathbf{u}_s$	Pixel index on Screen or its position in the world coordinate frame
l	Simple geometric mapping function
L	Geometric mapping function
$\mathcal{D}$	Deflectometric measurement. Consists of camera intrinsics $\mathcal{D}.\mathbf{A}$, camera extrinsics $\mathcal{D}.\check{\mathcal{E}}$, screen $\mathcal{D}.\mathcal{S}$ and a simple geometric mapping function $\mathcal{D}.l$.
$\mathbf{E}_3$	3×3 identity matrix
$\arg(.)$	Argument of a complex number
$a \circ b$	Composition of functions: $(a \circ b)(x) = a(b(x))$
$\Omega(a, b)$	Mapping to polar coordinates
$\propto$	Proportionality
$\cup$	Union
$\cap$	Intersection

Chapter 1

Mathematical Foundations

Before we dive into the deflectometry and its inner workings, we need a few mathematical foundations. This chapter will provide the basis for the geometric representations and calculations required for the solution of the deflectometric tasks we will encounter, especially the surface reconstruction algorithms. We will start out by introducing some lesser known number systems which we use to express three-dimensional geometry in an elegant fashion. We then provide mathematical representations for the objects we deal with in this work.

1.1 Number systems

We will start out by describing some number systems that are not generally well known – though they should be. These are namely the dual numbers and the quaternions and also their combinations – the dual Quaternions. These number systems will be introduced formally and their important properties will be discussed subsequently.

1.1.1 Dual Numbers

Dual numbers are a form of hypercomplex numbers. They were introduced by Clifford [Cli71] and further developed by Sturdy [Stu91] in the 18th century. They were mostly of mathematical interest for a long time but are more in use nowadays because of their ease of use in geometrical contexts. Our introduction is more modern and follows loosely [Dan99].

We will present them in analogy to the complex numbers which they closely resemble. We start out by introducing a new element ϵ into the real numbers. Another valid formulation can be found by identifying them with special members from $\mathbb{R}^2$.

We define a dual number $\breve{a}$ as a tuple of two real numbers and write

$$\breve{a} = a_r + \epsilon\, a_d \quad \text{with } a_r, a_d \in \mathbb{R}. \tag{1.1}$$

In analogy to i for the complex numbers, ϵ is a new base element. Unlike i, ϵ is nilpotent. That is

$$\epsilon^2 = 0. \tag{1.2}$$

The addition and multiplication known from the field of the real numbers make the dual numbers to an Abelian ring often named Δ. It is not a field because there is no inverse element for all $\breve{a} \in \Delta$ with $a_r = 0$.

The product of two dual numbers $\breve{a}$ and $\breve{b}$ reads

$$(a_r + \epsilon\, a_d)(b_r + \epsilon\, b_d) = a_r b_r + \epsilon\, (a_r b_d + a_d b_r). \tag{1.3}$$

Note that we keep the order of elements here because as soon as we use dual constructions over non-commutative algebras it will become important.

One very convenient and important aspect of dual numbers becomes clear when we look at a function of duals, i.e. $f(a + \epsilon\, b)$.

Let's take the Taylor expansion around the real point a.

$$f(a + \epsilon\, b) = \sum_{i=0}^{\infty} \frac{f^{(n)}(a)}{n!}(a + \epsilon\, b - a)^n \tag{1.4}$$

$$= f(a) + \epsilon\, b f'(a) + \frac{f''(a)}{2}(\epsilon\, b)^2 + \frac{f'''(a)}{6}(\epsilon\, b)^3 + \ldots \tag{1.5}$$

$$= f(a) + \epsilon\, b f'(a). \tag{1.6}$$

Of course, vectors of dimension n can also be constructed from Δ^n. Of special interest in the world of computational geometry – and in this work – are the elements of Δ^3. For example a three-dimensional vector $\check{x} \in \Delta^3 = x_r + \epsilon\, x_d$ with $x_r \cdot x_d = 0$ is a representation of a line in three dimensions as we will see in section 1.2.3.

1.1.2 Quaternions

Hamilton's work to extend complex numbers to higher dimensions led to the quaternions. Like for complex numbers, there are plenty of equivalent formulations for quaternions with different advantages and shortcomings. We will introduce two formulations which will be both used whenever they are convenient. A very good and more complete discussion of quaternions can be found in [Kui99].

The first formulation follows the analogy of the complex numbers. Starting from the base elements 1 and i known from the complex numbers, add two new base elements j and k such that

$$i^2 = j^2 = k^2 = ijk = -1. \tag{1.7}$$

Note that i, j and k are non-commutative – but associative – elements, i.e.

$$(ijk)k = -1k \Rightarrow \qquad ij(-1) = -k \Rightarrow \qquad ij = k \tag{1.8}$$

$$(ji)(ijk) = -(ji) \Rightarrow \qquad (-1)^2 k = -(ji) \Rightarrow \qquad ji = -k \tag{1.9}$$

The calculation of quaternion now follows naturally from the calculation with real numbers, i.e the product of two quaternions is

$$
\begin{aligned}
(a_s + a_x i + a_y j + a_z k)(b_s + b_x i + b_y j + b_z k) = \\
a_s b_s - (a_x b_x + a_y b_y + a_z b_z) \\
+i\,(a_s b_x + b_s a_x + a_y b_z - a_z b_y) \\
+j\,(a_s b_y + b_s a_y + a_z b_x - a_x b_z) \\
+k\,(a_s b_z + b_s a_z + a_x b_y - a_y b_x)
\end{aligned}
\tag{1.10}
$$

The second formulation is less intuitive but often more practical. We introduce the quaternion as a tuple $\mathfrak{a} = (a_s, \mathbf{a})$ with $\mathbf{a} = (a_x, a_y, a_z)^\mathsf{T}$ consisting of a scalar $a_s = \mathrm{Sc}\,(\mathfrak{a}) \in \mathbb{R}$ and a vector $\mathbf{a} = \mathrm{Vec}\,(\mathfrak{a}) \in \mathbb{R}^3$. We define the addition and scalar multiplication

$$
\mathfrak{a} + \mathfrak{b} = (a_s + b_s, \mathbf{a} + \mathbf{b}),
\tag{1.11}
$$

$$
\lambda \mathfrak{a} = (\lambda a_s, \lambda \mathbf{a}) \qquad \text{with } \lambda \in \mathbb{R}.
\tag{1.12}
$$

These operations make the quaternions a vector space over the real numbers which is generally named $\mathcal{H}$. As we will see below, there is a natural way to introduce a norm which makes $\mathcal{H}$ even to a normed vector space. The quaternion product (1.10) is now introduced as a new operation

$$
\mathfrak{a}\mathfrak{b} = (a_s b_s - \mathbf{a}^\mathsf{T} \mathbf{b},\, a_s \mathbf{b} + b_s \mathbf{a} + \mathbf{a} \times \mathbf{b}).
\tag{1.13}
$$

As

$$
(1,0)\mathfrak{a} = \mathfrak{a} = \mathfrak{a}(1,0),
\tag{1.14}
$$

this operation has a neutral element. It is also associative, but because of the cross product in the vector part it is not commutative.

Now, we ask the question in which cases the quaternion product can be zero. For this to be true, the following conditions must be met

$$
a_s b_s - \mathbf{a} \cdot \mathbf{b} = 0
\tag{1.15}
$$

$$
a_s \mathbf{b} + b_s \mathbf{a} + \mathbf{a} \times \mathbf{b} = \mathbf{0}.
\tag{1.16}
$$

Multiplying the first equation with b_s and the second with $\mathbf{b}$ and adding them up gives

$$0 = b_s(a_s b_s - \mathbf{a} \cdot \mathbf{b}) + \mathbf{b} \cdot (a_s \mathbf{b} + b_s \mathbf{a} + \mathbf{a} \times \mathbf{b}) \qquad (1.17)$$

$$= a_s b_s^2 + a_s \|\mathbf{b}\|^2 \qquad (1.18)$$

$$= a_s (b_s^2 + \|\mathbf{b}\|^2) \qquad (1.19)$$

So either $\mathbf{b} = \mathbf{o}$ or $a_s = 0$. But if a_s is zero, (1.16) becomes

$$b_s \mathbf{a} = \mathbf{b} \times \mathbf{a}. \qquad (1.20)$$

This says that $\mathbf{a}$ must be collinear to a vector perpendicular to $\mathbf{b}$ and $\mathbf{a}$ which is only possible if $\mathbf{a} = \mathbf{0}$. Therefore we've shown that the quaternions are zero divisor free. This makes $\mathcal{H}$ with the quaternion product to an associative division algebra.

We use the definition of the quaternion product to also introduce a product between quaternion $\mathbf{q} \in \mathcal{H}$ and vector $\mathbf{v} \in \mathbb{R}^3$

$$M : \mathcal{H} \times \mathbb{R}^3 \mapsto \mathbb{R}^3 \qquad (1.21)$$

$$\mathbf{q}\mathbf{v} \to \mathrm{Vec}\left(\mathbf{q}(0, \mathbf{v})\right). \qquad (1.22)$$

The product from the right side can be defined analogously. This operation is very useful when we discuss the quaternions in the sense of rotations in section 1.3.4. Note that this operation can be made implicit if we say that a vector $\mathbf{v}$ can be augmented to a quaternion $\mathbf{v} = (0, \mathbf{v})$. In fact we will introduce more augmentations for other constructs later.

In some cases, it will become useful to write quaternions in a linear fashion so that the common tools known from linear algebra can be applied to them. Very similar to how the cross product $[\mathbf{x}]_\times$ can be written as a matrix that can later be right multiplied with a vector so can the quaternion product be rewritten as

$$\mathbf{a}\mathbf{b} = \mathbf{L}(\mathbf{a})\mathbf{b} = \mathbf{R}(\mathbf{b})\mathbf{a} \qquad (1.23)$$

with

$$\mathbf{L}(\mathbf{a}) = \begin{pmatrix} a_s & -a_x & -a_y & -a_z \\ a_x & a_s & -a_z & a_y \\ a_y & a_z & a_s & -a_x \\ a_z & -a_y & a_x & a_s \end{pmatrix} \text{ and } \mathbf{R}(\mathbf{b}) = \begin{pmatrix} b_s & -b_x & -b_y & -b_z \\ b_x & b_s & b_z & -b_y \\ b_y & -b_z & b_s & b_x \\ b_z & b_y & -b_x & b_s \end{pmatrix}.$$

$$(1.24)$$

The quaternion to be multiplied is now seen as a vector from $\mathbb{R}^4$. Care must be taken in the order of scalar and vector parts in the vector.

We finish the introduction of quaternions by defining some common operations that are analogous to complex numbers. The conjugate of a quaternion $\mathbf{a} = (a_s, \mathbf{a})$ is defined as

$$\overline{\mathbf{a}} := (a_s, -\mathbf{a}). \tag{1.25}$$

The quaternion norm is

$$\|\mathbf{a}\| = \sqrt{\mathbf{a}\overline{\mathbf{a}}} \tag{1.26}$$

which is always a positive real number and fulfills all the axioms of a norm. Note that we are implicitly converting form a quaternion with no vector part to a scalar as the square root of a quaternion is not clearly defined.

1.1.3 Dual Quaternions

Dual quaternions – or bi-quaternions – follow naturally from the definitions of dual numbers and quaternions. They provide an intuitive representation of movements in three dimensions and have been introduced by Clifford as the first example of a new type of associative algebra which now carries his name [Cli71]. A somewhat more applied and modern approach to this topic can be found in [BR79].

A dual quaternion $\breve{\mathbf{a}} = (\breve{a}_s, \breve{\mathbf{a}})$ consists of a dual number $\breve{a}_s = \mathrm{Sc}\,(\breve{\mathbf{a}}) \in \Delta$ and a dual vector $\breve{\mathbf{a}} = \mathrm{Vec}\,(\breve{\mathbf{a}}) \in \Delta^3$. The three basic operations addition, multiplication and dual quaternion multiplication follow naturally from the corresponding operations of the dual numbers and the quaternions:

$$\breve{a} + \breve{b} = (\breve{a}_s + \breve{b}_s, \breve{\mathbf{a}} + \breve{\mathbf{b}}) \tag{1.27}$$

$$\lambda \breve{a} = (\lambda \breve{a}_s, \lambda \breve{\mathbf{a}}) \qquad \text{with } \lambda \in \mathbb{R} \tag{1.28}$$

$$\breve{a}\breve{b} = (\breve{a}_s \breve{b}_s - \breve{\mathbf{a}} \cdot \breve{\mathbf{b}}, \breve{a}_s \breve{\mathbf{b}} + \breve{b}_s \breve{\mathbf{a}} + \breve{\mathbf{a}} \times \breve{\mathbf{b}}) \tag{1.29}$$

Addition (1.27) and scalar multiplication (1.28) make the dual quaternions a module over Δ. Addition (1.27) and dual quaternion multiplication (1.29) make them a non Abelian ring. All operations taken together make them an associative algebra.

Note that both ways of looking at dual quaternions are valid: either one interprets them as a dual construct consisting of two quaternions – a real one and a dual one – or, as they have been introduced here, one can interpret them as a quaternion of a dual number and a dual vector. Both ways are useful in different situations and of course, both are equivalent as

$$\breve{a} = (\breve{a}_s, \breve{\mathbf{a}}) = (a_{sr} + \epsilon\, a_{sd}, \mathbf{a}_r + \epsilon\, \mathbf{a}_d) = (a_{sr}, \mathbf{a}_r) + \epsilon\, (a_{sd}, \mathbf{a}_d). \tag{1.30}$$

Here, we used the addition operation of dual numbers and the addition operation of quaternions.

There is some ambiguity in defining the conjugate of a dual quaternion. One could simply apply the rules of conjugation on sums and indeed this will be our first definition:

$$\overline{\breve{a}} := \overline{a_r + \epsilon\, a_d} = \overline{a}_r + \epsilon\, \overline{a}_d. \tag{1.31}$$

Another way could be to use a conjugation similar to the complex numbers where we change the sign in front of the second base element. This second type of conjugation is useful when using dual quaternions as operators acting on points in space (see section 1.4.2). We call it complete conjugation and denote it with two over bars:

$$\overline{\overline{\breve{a}}} := \overline{a}_r - \epsilon\, \overline{a}_d. \tag{1.32}$$

The norm of a dual number is a dual number with positive real part defined as

$$\|\breve{\mathbf{a}}\| = \sqrt{\breve{\mathbf{a}}\overline{\breve{\mathbf{a}}}} = \sqrt{(\mathbf{a}_r + \epsilon\,\mathbf{a}_d)(\overline{\mathbf{a}}_r + \epsilon\,\overline{\mathbf{a}}_d)} \tag{1.33}$$

$$= \sqrt{\mathbf{a}_r\overline{\mathbf{a}}_r + \epsilon\,(\mathbf{a}_r\overline{\mathbf{a}}_d + \mathbf{a}_d\overline{\mathbf{a}}_r)} \tag{1.34}$$

$$= \sqrt{\mathbf{a}_r\overline{\mathbf{a}}_r} + \epsilon\,\frac{\mathbf{a}_r\overline{\mathbf{a}}_d + \mathbf{a}_d\overline{\mathbf{a}}_r}{2\sqrt{\mathbf{a}_r\overline{\mathbf{a}}_r}} \in \Delta. \tag{1.35}$$

If the real part is nonzero, the dual quaternion has an inverse which is simply

$$\breve{\mathbf{a}}^{-1} = \|\breve{\mathbf{a}}\|^{-1}\overline{\breve{\mathbf{a}}}. \tag{1.36}$$

1.2 Geometric Representation

We will now briefly discuss the representation of various geometric elements and operations used in this work.

1.2.1 Points

The basic representation of points in this work are three tuples in a vector notation

$$\mathbf{p} = (x, y, z)^\mathsf{T}. \tag{1.37}$$

We will augment this notation transparently when appropriate either to work in homogeneous coordinates or when we are transforming a point through a rigid mapping using the corresponding dual quaternion (see section 1.4). The bijective transformations used for these two operations are

$$\mathbb{R}^3 \mapsto \mathbb{R}^4 \tag{1.38}$$

$$(x, y, z)^\mathsf{T} \to (x, y, z, 1)^\mathsf{T} \tag{1.39}$$

and

$$\mathbb{R}^3 \mapsto \breve{\mathcal{H}} \tag{1.40}$$

$$(x, y, z)^\mathsf{T} \to (1, 0, 0, 0) + \epsilon\,(0, x, y, z) \tag{1.41}$$

The last transformation might seem unusual, but will become useful when we introduce the geometric sandwich product for point transformations in section 1.4.2.

1.2.2 Planes

Planes do not fit into our framework as nicely as the other geometric entities we work with. In this work, we represent planes as a quaternion, whereas the scalar part is the distance from the origin and the vector part is the normal vector. This stretches the interpretation of quaternions over its limits and is more of an implementation clutch than theoretically warranted.

The quaternion $\mathbf{p} = (d, x_0, y_0, z_0)$ represents the plane

$$(x_0, y_0, z_0) \cdot \mathbf{x} + d = 0. \tag{1.42}$$

Note that this representation cannot be rotated using quaternions or transformed using dual quaternions without loosing its inner geometry. Therefore those transformations are not physically correct and therefore considered wrong for this work. The complete framework of Geometric Algebra [DF07] – from which our number system representations are a subset – offers a beautiful representation of planes using bivectors which keep all of these transformational properties. We refrained from using the complete framework of Geometric Algebra because using it would have made the mathematics for this work require too much background reading. It is worth studying though.

1.2.3 Lines

A line is fully represented by two points in space through which it passes. Alternatively, one can use a point and a direction. Both choices are rather arbitrary: there are infinitively many equivalent choices for each line. Some of them might be better suited for numerical calculations than others. For this work, we choose a representation that is unique for each (directed) line, numerically stable and ties in well with the quaternions and dual quaternions used to represent rotations and rigid transformations: normalized Plücker coordinates [Gal01].

We define a directed line from point $\mathbf{p}_1$ to $\mathbf{p}_2$ as the dual vector

$$\check{\mathbf{l}} = (\mathbf{d} + \epsilon\, \mathbf{m})/\|\mathbf{d}\| \tag{1.43}$$

with $\mathbf{d} = \mathbf{p}_2 - \mathbf{p}_1$ being the direction of the line and $\mathbf{m} = \mathbf{p}_2 \times \mathbf{p}_1$ being the moment. The name moment is chosen because if a unit mass were to move from $\mathbf{p}_1$ to $\mathbf{p}_2$ it would have $\mathbf{m}$ as its moment. This representation is normalized because we force the direction to be a unit vector. Also note that the dual and the real part of the vector are orthogonal to each other. Therefore this representation has 4 degrees of freedom just like a line in space. This makes the representation unique.

Given two such lines $\check{\mathbf{x}}$ and $\check{\mathbf{y}}$ with the additional constraint that they are normalized, that is $\|\mathbf{x}_r\| = \|\mathbf{y}_r\| = 1$, the dot product

$$\cos\check{\Theta} = \check{\mathbf{x}} \cdot \check{\mathbf{y}} = (\mathbf{x}_r \cdot \mathbf{y}_r) + \epsilon\,(\mathbf{x}_r \cdot \mathbf{y}_d + \mathbf{x}_d \cdot \mathbf{y}_r) \tag{1.44}$$

is equal to the cosine of a dual angle $\check{\Theta} = \Theta + \epsilon\, d$. Hereby the real part is the angle between the lines while d is their smallest distance. This property is derived in section A.1.

Of importance for us is also the plane-line meet operation, that is the point of intersection $\mathbf{p}$ between a plane $\mathbf{p} = (d, \mathbf{n})$ and a line not inside this plane $\check{\mathbf{l}} = \mathbf{l} + \epsilon\, \mathbf{m}$. This point can be calculated using the relation

$$\mathbf{p} = \frac{\mathbf{m} \times \mathbf{n} - d\, \mathbf{l}}{\mathbf{n} \cdot \mathbf{l}} \tag{1.45}$$

which is derived in section A.2.

Plücker coordinates tie in nicely with the dual quaternion's representation: augmenting a dual vector into a dual quaternion is done equivalently to how a vector is turned into a quaternion: keep the vector part and set the scalar part to zero. This dual quaternion can then be transformed using the dual quaternion product. This gives an easy and elegant way to transform lines through rigid transformations (see section 1.4.2).

1.2.4 Free Form Surfaces

The gist of this work is on the reconstruction of free form surfaces representable as a function on a Cartesian grid

$$z : \mathbb{R}^2 \mapsto \mathbb{R} \tag{1.46}$$

$$(x, y) \to z(x, y). \tag{1.47}$$

Therefore, we represent surfaces as point clouds or – when appropriate – as a discrete set of $\{x, y\}$ values (e.g. camera pixel coordinates) and how they relate to each other in neighbourship and the corresponding $z(x, y)$ value.

There are plenty of alternative models available to represent free form surfaces under some smoothness constraints. These are important for a proper interpolation of the raw point-cloud data or for data reduction. In the simulations for this work we used NURBS [Far99] as the most flexible representation – this work is not concerned with the interpolation of point clouds using NURBS though.

1.3 Representation of Rotations

Rotations in three dimensions have three degrees of freedom. Because of this, most people immediately relate them to Euler angles - i.e. three angles around the axis of a Cartesian coordinate system. The order in which the rotations are applied has to be defined and this representation has problems with extreme angles which can easily occur in our case. The shortcomings will be discussed below in section 1.3.2.

Another approach is to use rotation matrices. They no longer contain the ambiguity of Euler angles, but they contain a lot of redundant information: nine entries instead of just three.

A more elegant approach is to use Rodrigues' rotation formula which describes a rotation around a rotation axis $\mathbf{k} \in \mathbb{R}^3$ by an angle Φ. Therefore we have three real values for three degrees of freedom, the most compact representation we can hope for. However, this representation cannot be used for efficient and numerically stable calculations.

Finally, this leads to the quaternion representation of rotations. Unit length quaternions contain the same information as the Rodrigues formula but offer a higher numerical stability in evaluation and more mathematical expression power. For these reasons, quaternions were used for most implementations in this work – but the other representations are sometimes easier to work with in a theoretical context.

1.3.1 Rotation Matrices

The rotation matrices in three dimensions are the members of the special orthogonal group $SO(3)$ which forms a subgroup in the field of $\mathbb{R}^{3\times 3}$. The properties of any $\mathbf{R} \in SO(3)$ are

$$\mathbf{R}^{-1} = \mathbf{R}^{\mathsf{T}} \quad \text{and} \quad \det(\mathbf{R}) = 1. \tag{1.48}$$

Now, assume a rotation around an axis $\mathbf{n}$ which passes through the origin by an angle of Φ. Its rotation matrix respective to a fixed base of $\mathbb{R}^3$ is called $\mathbf{R}$. Any vector $\mathbf{v} = k\mathbf{n}$ with $k \in \mathbb{R}$ will be collinear to the rotational axis $\mathbf{n}$ and will therefore not change under the given rotation:

$$\mathbf{R}\mathbf{v} = \mathbf{v}. \tag{1.49}$$

This means that any vector along the axis of rotation is an eigenvector of $\mathbf{R}$ to the eigenvalue 1. Specifying this even further, there is always another element $\mathbf{R}_v \in SO(3)$ which is conjugate to $\mathbf{R}$

$$\mathbf{R}_v = \mathbf{S}\mathbf{R}\mathbf{S}^{\mathsf{T}} \quad \text{with an appropriate } \mathbf{S} \in SO(3) \tag{1.50}$$

which expresses this rotation in the simple form of

$$\mathbf{R}_v = \begin{pmatrix} 1 & 0 & 0 \\ 0 & \cos(\Phi) & -\sin(\Phi) \\ 0 & \sin(\Phi) & \cos(\Phi) \end{pmatrix}. \tag{1.51}$$

This makes the other two eigenvalues $e^{\pm i\Phi}$ apparent.

To rotate a vector $\mathbf{v} \in \mathbb{R}^3$ with the rotation matrix $\mathbf{R}$, one simply multiplies $\mathbf{v}$ from the right with $\mathbf{R}$:

$$\mathbf{v}_{\mathrm{rot}} = \mathbf{R}\mathbf{v} \tag{1.52}$$

This is the cheapest way to rotate a vector requiring nine multiplications and six additions. However, when one wants to compose two rotations $\mathbf{R}_1$ and $\mathbf{R}_2$, the additional matrix multiplication increases the cost dramatically. The combined transformation

$$\mathbf{R}_1\mathbf{R}_2 \tag{1.53}$$

takes 27 multiplications and 18 additions.

1.3.2 Euler Angles

To specify a rotation in three dimensions with three rotation angles one needs to agree around which axis the rotations should be done. The term Euler angles implies for some sources already a certain choice, but in this work we will use the term more generally. Nevertheless for practical applications, the axis must be defined and fixed. We use the convention of the robot which is part of our experimental setup. Given three orthogonal axis $\mathbf{e}_x$, $\mathbf{e}_y$ and $\mathbf{e}_z$ and three angles A, B and C, the rotation is first done around the z-axis $\mathbf{e}_z$ by the angle A, then around the new y-axis $\mathbf{e}_y'$ by the angle B and lastly around the new x-axis $\mathbf{e}_x''$ by the angle C. This representation is sometimes called improper Euler angles or Tait-Bryan angles. It is also the DIN 9300 norm used in aeronautics where it is called Yaw-Pitch-Roll.

Constructing a rotation matrix is now tedious, but easy: We simply multiply the three rotation matrices:

$$\mathbf{R}(A, B, C) = \mathbf{R}_x(C)\mathbf{R}_y(B)\mathbf{R}_z(A) =$$

$$\begin{pmatrix} 1 & 0 & 0 \\ 0 & \cos(C) & \sin(C) \\ 0 & -\sin(C) & \cos(C) \end{pmatrix} \begin{pmatrix} \cos(B) & 0 & -\sin(B) \\ 0 & 1 & 0 \\ \sin(B) & 0 & \cos(B) \end{pmatrix} \begin{pmatrix} \cos(A) & \sin(A) & 0 \\ -\sin(A) & \cos(A) & 0 \\ 0 & 0 & 1 \end{pmatrix}$$

$$\begin{pmatrix} \cos(A)\cos(B) & \sin(A)\cos(B) & -\sin(B) \\ -\sin(A)\cos(C)+\sin(B)\sin(C)\cos(A) & \sin(A)\sin(B)\sin(C)+\cos(A)\cos(C) & \sin(C)\cos(B) \\ \sin(A)\sin(C)+\sin(B)\cos(A)\cos(C) & \sin(A)\sin(B)\cos(C)-\sin(C)\cos(A) & \cos(B)\cos(C) \end{pmatrix}$$

$$\tag{1.54}$$

One fundamental problem of Euler angles besides the confusion related to the choice of rotation axis is that it encodes what is essentially one rotation as three consecutive rotations. This problem becomes apparent if one chooses $B = \pm\pi/2$ in (1.54):

$$\mathbf{R}(A, \pm\pi/2, C) = \begin{pmatrix} 0 & 0 & \mp 1 \\ \pm\sin(C\mp A) & \cos(C\mp A) & 0 \\ \pm\cos(C\mp A) & -\sin(C\mp A) & 0 \end{pmatrix} \qquad (1.55)$$

Now, there is no algebraic difference between changes in A or C and therefore they can no longer be used to differentiate between different rotational axis. This phenomenon is called Gimbal lock. Another weak point of the Euler angles is the asymmetry between rotations and their inverse. For example, the rotation $(89.9°, 0, 90.1°)$ which „nearly" changes $x \to y$, $y \to z$ and $z \to x$ is inverted by the rotation $(45°, -89.9°, -135°)$. There is also no immediate way to rotate a vector using only the Euler angles. One needs to convert to one of the other representations – i.e. the corresponding rotation matrix – before the actual rotated vector can be calculated.

1.3.3 Rodrigues' Rotation Formula

The idea to represent a rotation via an axis and a corresponding rotation angle Φ is very basic. The Rodrigues parameters encode this information in one vector $\mathbf{n}$, such that the length of $\mathbf{n}$ represents the angle $\Phi = \|\mathbf{n}\|$ and the direction of $\mathbf{n}$ represents the rotational axis $\mathbf{r} = \mathbf{n}/\|\mathbf{n}\| = \mathbf{n}/\Phi$. A rotation can then be performed via Rodrigues' rotation formula

$$\mathbf{v}_{\text{rot}} = \mathbf{v}\cos\Phi + (\mathbf{r}\times\mathbf{v})\sin\Phi + \mathbf{r}(\mathbf{r}\cdot\mathbf{v})(1-\cos\Phi). \qquad (1.56)$$

Note that this is essentially the same as calculating a rotation matrix

$$\mathbf{R} = \cos\Phi\,\mathbf{E}_3 + (1-\cos\Phi)\mathbf{r}\mathbf{r}^{\mathsf{T}} + \sin\Phi\,[\mathbf{r}]_\times \qquad (1.57)$$

and then right multiplying $\mathbf{v}$ with it. Here we used $\mathbf{E}_3$ as the identity matrix and $[\mathbf{r}]_\times$ is the matrix operator that represents the cross product

$$[\mathbf{r}]_\times\mathbf{v} = \mathbf{r}\times\mathbf{v} = \begin{pmatrix} 0 & -r_3 & r_2 \\ r_3 & 0 & -r_1 \\ -r_2 & r_1 & 0 \end{pmatrix}\mathbf{v}. \qquad (1.58)$$

The Rodrigues' formulation is very space efficient as it only needs three real numbers to represent a rotation in three-dimensional space. It is also intuitive to understand and visualize and it has no singularities. However, it has plenty of practical disadvantages: direct rotation of a vector is expensive, especially because the cosine and the sine of the angle have to be calculated but also because plenty of multiplications (23) and additions (16) must be done. There is also no easy way to compose two rotations using this representations without converting to any other representation.

1.3.4 Unit Quaternions

Quaternions have four degrees of freedom. When we restrict ourselves to unit length quaternions – sometimes also called versors, a designation we will avoid in this work – we are left with three degrees of freedom. This is the same number of degrees of freedom as Rodrigues' representation of rotations. In fact, we can find an easy mapping: given a Rodrigues' rotation vector $\mathbf{n}$, we can construct a quaternion $\mathbf{q}$ as

$$\mathbf{q} = (\cos(\Phi/2), \mathbf{l}\sin(\Phi/2)) \tag{1.59}$$

with $\Phi = \|\mathbf{n}\|$ and $\mathbf{l} = \mathbf{n}/\|\mathbf{n}\|$. This mapping is in fact one-to-one, therefore this is also a representation of a rotation. More precisely, the unit length quaternions are also a double-cover of the special orthogonal group SO(3) – therefore any rotation is represented by two unit length quaternions which differ only in sign. This is necessary for a representation of rotations to be continuous and without singularities. The reason for this is that the set of rotations in three dimensions is non orientable. The double cover avoids singularities and allows for easy interpolation of rotations [Gal01].

A rotation of a vector $\mathbf{v} = (0, \mathbf{v})$ can be done by the geometric sandwich product.

$$\mathbf{v}_{\text{rot}} = \mathbf{q}\mathbf{v}\overline{\mathbf{q}} \tag{1.60}$$
$$= (\cos\varphi, \mathbf{l}\sin\varphi)(0, \mathbf{v})(\cos\varphi, -\mathbf{l}\sin\varphi) \tag{1.61}$$
$$= (-\sin\varphi(\mathbf{l}\cdot\mathbf{v}), \cos\varphi\mathbf{v} + \sin\varphi\mathbf{l}\times\mathbf{v})(\cos\varphi, -\mathbf{l}\sin\varphi) \tag{1.62}$$

$$= (0, \sin^2 \varphi (\mathbf{l} \cdot \mathbf{v})\mathbf{l} \tag{1.63}$$

$$+ 2\sin\varphi\cos\varphi(\mathbf{l} \times \mathbf{v}) + \sin^2\varphi\mathbf{l} \times (\mathbf{l} \times \mathbf{v}) + \cos^2\varphi\mathbf{v} \tag{1.64}$$

$$= (0, (1 - \cos\Phi)(\mathbf{l} \cdot \mathbf{v})\mathbf{l} + \sin\Phi\mathbf{l} \times \mathbf{v} + \cos\Phi\mathbf{v}) \tag{1.65}$$

The Rodrigues' rotation formula (1.56) appears naturally in this operation. We introduced $\varphi := \Phi/2$, used that $\mathbf{l}$ has length 1 and used the following identities in this calculation:

$$\mathbf{l} \times (\mathbf{l} \times \mathbf{p}) = (\mathbf{l} \cdot \mathbf{v})\mathbf{l} - (\mathbf{l} \cdot \mathbf{l})\mathbf{v} = (\mathbf{l} \cdot \mathbf{v})\mathbf{l} - \mathbf{v} \tag{1.66}$$

$$\cos^2\varphi - \sin^2\varphi = \cos\Phi \tag{1.67}$$

$$2\sin^2\varphi = (1 - \cos\Phi) \tag{1.68}$$

$$2\sin\varphi\cos\varphi = \sin\Phi \tag{1.69}$$

The name of the operation comes from the quaternion $\mathbf{q}$ *sandwiching* the object to be transformed by multiplying it from left and from right. We will meet similar products in later sections again.

1.4 Representation of Rigid Transformations

We will need to work in many different frames of reference and we will also optimize transformations between various frames of reference. We therefore need a suitable representation of a frame-to-frame transformation. Since we always work in Cartesian coordinates, we can restrict us to a very special subset of affine transformations – namely rigid transformations. These preserve the local geometric properties and angles of transfered lines.

1.4.1 Linear Embedding through Homogeneous Coordinates

It is well known that each rigid motion can be represented by a rotation followed by a translation. Using homogeneous coordinates, rigid transformations can be embedded in a linear space. A transformation of a point **p** by the rotation represented by the 3×3 matrix **R** and the translation **t** can than be written as

$$\begin{pmatrix} \mathbf{p}_{tr} \\ 1 \end{pmatrix} = \begin{pmatrix} \mathbf{R} & \mathbf{t} \\ \mathbf{0} & 1 \end{pmatrix} \begin{pmatrix} \mathbf{p} \\ 1 \end{pmatrix}. \tag{1.70}$$

The complete transformation is therefore captured in the given 4×4 matrix. The inverse transformation is also captured through the inverse matrix. Composing two rigid motions is done by multiplying the corresponding matrices.

This representation however has the same flaws as matrices for the representation of rotations: it wastes space, composing is expensive and can become numerically instable. Also the concept of rotating a line is not straight forward to express.

1.4.2 Unit Dual Quaternions

Analogously to how we represent lines via a dual vector and rotations via a unit quaternion, we can also express rigid transformations using a unit length dual quaternion. The rotational part is already discussed: we use one of the real unit length quaternion that corresponds to the rotation matrix **R**. This will be the real part $\mathbf{q}_r$ of our dual quaternion $\breve{\mathbf{q}}$. For the translation part represented through $\mathbf{t} = (t_x, t_y, t_z)^\mathsf{T}$ we define

$$\mathbf{q}_d := \frac{1}{2}(0, t_x, t_y, t_z)\mathbf{q}_r = \frac{1}{2}\mathbf{t}\mathbf{q}_r \tag{1.71}$$

using the quaternion $\mathbf{t} = (0, \mathbf{t})$. This gives us the dual quaternion

$$\breve{\mathbf{q}} = \mathbf{q}_r + \epsilon\,\mathbf{q}_d \tag{1.72}$$

with the squared norm

$$\|\breve{\mathbf{q}}\|^2 = \mathbf{q}_r\overline{\mathbf{q}}_r + \frac{\epsilon}{2}\left(\mathbf{t}\mathbf{q}_r\overline{\mathbf{q}_r} + \mathbf{q}_r\overline{(\mathbf{t}\mathbf{q}_r)}\right) \tag{1.73}$$

$$= 1 + \frac{\epsilon}{2}(\mathbf{t} + \overline{\mathbf{t}}) = 1. \tag{1.74}$$

This is therefore indeed a unit length dual quaternion. We can use this representation to transform a point $\mathbf{x}$ by first augmenting it to a dual quaternion $\check{\mathbf{p}} = 1 + \epsilon\,(0, \mathbf{x}) = 1 + \epsilon\,\mathbf{\underline{x}}$ and then sandwiching it between $\check{\mathbf{q}}$ and its complete conjugate $\overline{\overline{\check{\mathbf{q}}}}$.

$$\check{\mathbf{p}}_{\text{tr}} = \check{\mathbf{q}}\check{\mathbf{p}}\overline{\overline{\check{\mathbf{q}}}} \tag{1.75}$$

$$= (\mathbf{q}_r + \epsilon\,\mathbf{q}_d)(1 + \epsilon\,\mathbf{\underline{x}})(\overline{\mathbf{q}}_r - \epsilon\,\overline{\mathbf{q}}_d) \tag{1.76}$$

$$= \mathbf{q}_r\overline{\mathbf{q}}_r + \epsilon\,(\mathbf{q}_r\mathbf{\underline{x}}\overline{\mathbf{q}}_r + \mathbf{q}_d\overline{\mathbf{q}}_r - \mathbf{q}_r\overline{\mathbf{q}}_d) \tag{1.77}$$

$$= 1 + \epsilon\,\left(\mathbf{q}_r\mathbf{\underline{x}}\overline{\mathbf{q}}_r + \frac{1}{2}\mathbf{\underline{t}}\mathbf{q}_r\overline{\mathbf{q}}_r - \frac{1}{2}\mathbf{q}_r\overline{\mathbf{q}}_r\mathbf{\underline{t}}\right) \tag{1.78}$$

$$= 1 + \epsilon\,(0, \mathbf{Rx} + \mathbf{t}) \tag{1.79}$$

We can also compose rigid transformations or transform lines (represented by a dual vector augmented to a dual quaternion) similarly, but using only the normal conjugation

$$\check{\mathbf{l}}_{\text{tr}} = \check{\mathbf{q}}\check{\mathbf{l}}\overline{\check{\mathbf{q}}}. \tag{1.80}$$

Here again, we see variations of the sandwich product which we first met in section 1.3.4. Both operations read very similarly to how unit length quaternions are used to rotate vectors.

This representation shares the advantages of quaternions and many of its properties. There are two representations for each rigid transformation, they are easy to interpolate or change by a small value, they are efficient when composed and are numerically stable. However, more multiplications and additions are needed for a single transformation than for the corresponding matrix representation.

The true power of dual quaternions becomes apparent when we identify them with screw motions. To make this clear, we state the following theorem:

Chasles' Theorem. *The most general rigid displacement can be produced by a translation along a line followed (or preceded) by a rotation about that line. That is, any rigid displacement can be represented by a screw like movement. This relationship is visualized in Figure 1.1.*

Proof. We will proceed in two steps. In the first step, we will convert a transformation given as rotation matrix $\mathbf{R}$ and translation vector

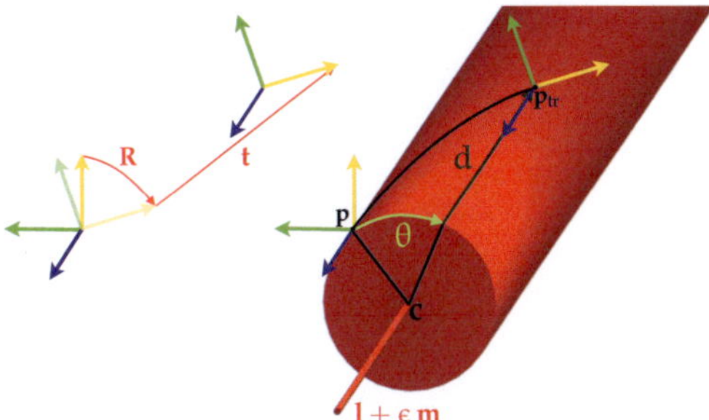

Figure 1.1: Comparison of representations for the same rigid movement. On the left, the coordinate system is first rotated, then translated. Order matters. On the right, the coordinate systems is rotated around and moved along a screw axis. Order does not matter – when both transformations are done at the same time, one gets a screw like motion.

$\mathbf{t}$ into a screw motion. In the second step, we will show that we can always find a dual quaternion $\breve{q}$, such that (1.75) is the same transformation than the original.

We will start by finding the screw. We need to determine the direction $\mathbf{l}$ and the moment $\mathbf{m}$ of the screw axis. We also need the angle of rotation Φ around the axis as well as the pitch d along it. The direction $\mathbf{l}$ and the angle Φ are already known, they remain the same for the screw motion. The pitch d is simply the projection of $\mathbf{t}$ on the direction of the screw $d = \mathbf{t} \cdot \mathbf{l}$. To find the moment $\mathbf{m}$, we consider one point $\mathbf{p}$ on the screw. If we knew $\mathbf{p}$, we could find the moment as $\mathbf{m} = \mathbf{l} \times \mathbf{p}$. We know that the transformation of $\mathbf{p}$ will also be on the screw but shifted by d along the axis:

$$\mathbf{p} = \mathbf{R}\mathbf{p} + \mathbf{t} - d\mathbf{l}. \tag{1.81}$$

We decompose the rotation matrix according to Cayley's formula using the Rodrigues' vector $\mathbf{b} = \mathbf{l}\tan\varphi$ with $\varphi := \Phi/2$.

$$\mathbf{p} = (\mathbf{E}_3 - [\mathbf{b}]_\times)^{-1}(\mathbf{E}_3 + [\mathbf{b}]_\times)\mathbf{p} + \mathbf{t} - d\mathbf{l}. \tag{1.82}$$

Multiplying with $(\mathbf{E}_3 - [\mathbf{b}]_\times)$ and rearranging gives

$$\frac{1}{2}(\mathbf{b} \times \mathbf{t} - \mathbf{t} + d\mathbf{l}) = \mathbf{b} \times \mathbf{p}. \tag{1.83}$$

We now take the cross product with $\mathbf{b}$ from the left and use that $\mathbf{p}$ and $\mathbf{b}$ are orthogonal and (1.66):

$$\mathbf{p} = \frac{\mathbf{b} \times \mathbf{t} - \mathbf{b} \times (\mathbf{b} \times \mathbf{t})}{2\mathbf{b} \cdot \mathbf{b}} \tag{1.84}$$

We proceed to construct $\mathbf{m}$ as

$$\mathbf{m} = \mathbf{l} \times \mathbf{p} = \frac{(\mathbf{l} \cdot \mathbf{t})\mathbf{l} - \mathbf{t} + \tan\varphi(\mathbf{l} \times \mathbf{t})}{2\tan\varphi}. \tag{1.85}$$

We will now construct a dual quaternion $\breve{\mathbf{q}}$ containing the information we derived. We represent the screw axis with a dual vector $\breve{\mathbf{l}} := \mathbf{l} + \epsilon\,\mathbf{m}$ and combine the rotation angle and the pitch in a dual scalar $\breve{\Phi} := \Phi + \epsilon\,d$, the dual quaternion representing the rigid transformation reads as

$$\breve{\mathbf{q}} = \big(\cos(\breve{\Phi}/2), \breve{\mathbf{l}}\sin(\breve{\Phi}/2)\big) \tag{1.86}$$

$$= (\cos(\Phi/2), \mathbf{l}\sin(\Phi/2)) + \tag{1.87}$$

$$\epsilon\big(-d/2\sin(\Phi/2), \mathbf{l}d/2\cos(\Phi/2) + \mathbf{m}\sin(\Phi/2)\big) \tag{1.88}$$

$$=: \mathbf{q}_r + \epsilon\,\mathbf{q}_d. \tag{1.89}$$

It is remarkable, how similar (1.86) looks compared to (1.59) and it indeed contains this very rotational quaternion as its real part. To conclude, we transform the original point using (1.75) to prove that this transformation is indeed the same as the one represented by $\mathbf{R}$ and $\mathbf{t}$.

$$\breve{\mathbf{x}}_{tr} = (\mathbf{q}_r + \epsilon\,\mathbf{q}_d)(1 + \epsilon\,\mathbf{x})(\overline{\mathbf{q}}_r - \epsilon\,\overline{\mathbf{q}}_d) \tag{1.90}$$

$$= \mathbf{q}_r\overline{\mathbf{q}}_r + \epsilon\,(\mathbf{q}_r\mathbf{x}\overline{\mathbf{q}}_r + \mathbf{q}_d\overline{\mathbf{q}}_r - \mathbf{q}_r\overline{\mathbf{q}}_d) \tag{1.91}$$

$$= 1 + \epsilon\,\big(0, \mathbf{Rx} + \mathbf{l}d - 2\mathbf{m}\sin\varphi\cos\varphi - 2\sin^2(\varphi)(\mathbf{l} \times \mathbf{m})\big) \tag{1.92}$$

$$= 1 + \epsilon\,\big(0, \mathbf{Rx} + \mathbf{l}d - \cos^2(\varphi)(\mathbf{l} \cdot \mathbf{tl} - \mathbf{t}) - \sin^2(\varphi)(\mathbf{l} \cdot \mathbf{tl} - \mathbf{t})\big) \tag{1.93}$$

$$= 1 + \epsilon\,(0, \mathbf{Rx} + \mathbf{t}) \tag{1.94}$$

$$\square$$

Dual quaternions are a very flexible representation: The geometric properties of the screw are readily available in the dual quaternion as can be seen in (1.86). The dual quaternion from a rotation matrix and a vector $\mathbf{t}$ can be found by first constructing the real part $\mathbf{q}_r$ as the quaternion corresponding to $\mathbf{R}$. The dual part is then constructed as $\mathbf{q}_d = \frac{1}{2}(0,\mathbf{t})\mathbf{q}$. If the original transformation vector is needed again, it can be found as

$$(0,\mathbf{t}) = 2\mathbf{q}_d\overline{\mathbf{q}}_r. \tag{1.95}$$

The conversion between screw representation and rotation-translation is now also easily done using the dual quaternions as the bridge.

Chapter 2

Introduction and Scope

We are now equipped with the math and the understanding we need to discuss the deflectometry per se. This chapter will give an overview of the concept and ideas of the deflectometry and formulate its primary equations. We will also discuss its limit and close the chapter with a problem statement that defines what we want to tackle in the rest of this thesis.

2.1 Introduction

Visual inspection of industrial parts is an important topic and has been in the focus of research for more than three decades now. Its benefits are usually self selling: the inspection volume is very flexible and can be changed easily by just changing position or lens of a camera, the measuring process is inherently parallel for a large area, and it is non-destructive because no physical contact is needed for the measurement. In addition, the deployment is often straightforward: most industrial materials are dull and reflect only diffusely which make them easy to light and see with a camera. And in most situations surroundings and lighting can be controlled to be ideal for a specific part of interest.

(a) Ferris wheel by Philip Lenz

(b) Skyscraper in Wellington, NZ

Figure 2.1: Examples for the deflectometric principle.

There are some border cases where visual inspection becomes hard: if the part is too bulky or cannot be moved for other reasons the acquisition can become more difficult. Another common problem is that some parts are not or only partially reflecting diffusely. In fact, some industrial parts are highly or even completely specular or transparent. Traditional approaches to visual inspection fail here – the object is simply invisible in a camera or will only reflect its surroundings. However, even then visual inspection is possible using deflectometry.

The deflectometry is a technique that measures the slope of a reflecting free form surface. Its working principle is easily understood by looking at the images in Figure 2.1. The Figure 2.1(a) shows the reflection of a ferris wheel in water. From the a priori knowledge that the bars in the wheel are indeed straight one can directly deduce that the water is not completely flat: there must be waves. However, it is hard to assess how strongly the water is perturbed from this picture alone. The same can be said about the windows in Figure 2.1(b): though they appear to be flat when seen as surfaces, the reflection of the environment proofs otherwise: straight lines are distorted which gives away the fact that the windows are indeed curved. The curvature is not visible by itself – not even from the inside of the house - which gives a clue on the sensitivity of this approach. The deflectometric principle is the formalization of this method: given a structured light source of known geometry, look at its distorted reflection in the object of interest and deduce the geometric properties of said object from this image or a series of images [RS10].

The deflectometry has been well established in recent years as a tool for the qualitative inspection of highly reflecting surfaces. In fact it has been shown that it can be a very sensitive method and detect smaller defects than most of the methods for diffusely reflecting parts [IH79]. However, the quantitative inspection – i.e. an inspection that also learns the depth of bumps or the length of scratches – is a much desired improvement. As of yet, it is not possible to acquire a precise and absolute three-dimensional reconstruction of a defect with the commercial systems available today. There are two reasons for this: first, quantitative reconstruction requires a calibration of the deflectometric system which is often either not possible or not feasible. The second reason is that the reconstruction of general three-dimensional surfaces from deflectometric measurements is not yet well understood and remains an active topic of research.

This work makes a contribution in the area of three-dimensional reconstruction of fully reflecting free form surfaces using deflectometry. It describes in detail various methods for the reconstruction of arbitrary free form surfaces from one or more deflectometric measurements. A suggestion is made to ease the needed calibration effort which makes it possible to move the structured light source and/or the camera without requiring a precise recalibration. All

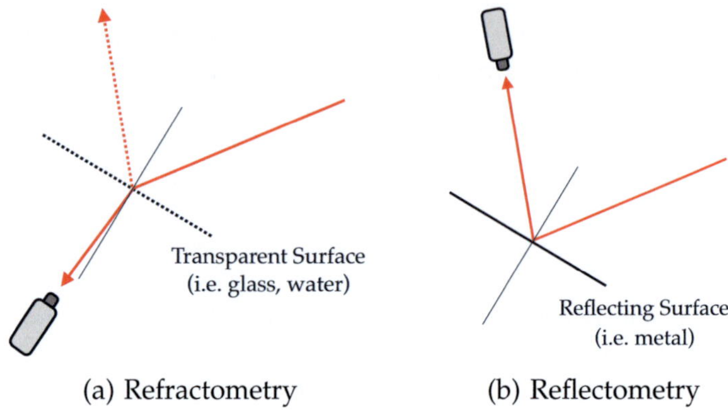

Figure 2.2: The two kinds of deflectometry.

methods are evaluated in simulations and experiments.

2.2 Types of Deflectometry

The Oxford Dictionary of English [Dic89] defines *to deflect* as a verb meaning to "cause (something) to change direction by interposing something; turn aside from a straight course". There are only two effects in non-relativistic physics which can impose a change of direction upon a light ray.

The first one is refraction – a light wave leaves one medium and enters another with a different index of refraction. In the new medium, the phase velocity of the wave is different which leads to a change of path of the beam following the d'Alembert principle [Dem]. Given a known structured light source and the possibility to capture the disturbed image of the light source after its light has traversed a transparent surface, some information about the normals of the transparent surface can be acquired. We call this specific version of deflectometry *refractometry*. It is depicted in Figure 2.2(a). The amount of change in the path is defined via Snell's Law. The change depends on the geometry of the setup, the refractive indices of the participating materials and the frequency of the light source. Refractometry has been applied to great success in reconstruction of water waves in small controlled indoor experiments [JSR05, JKW94].

The other version of deflectometry is *reflectometry*. It is conceptionally simpler because it has no dependency on the wave length of the light. Instead of the refracted image of the light source, one now captures its reflected and distorted image. Obviously the surface under investigation must now be specular reflective instead of transparent. The direction of the reflected beam is then a function of the surface normal and the direction of the incoming beam. The basic principle of reflectometry is depicted in Figure 2.2(b).

The term *deflectometry* is used in the literature synonymously with the process we just introduced as reflectometry. Though imprecise and confusing with respect to refractometry we will follow suite and drop the distinction in terms between deflectometry and reflectometry and will only talk about deflectometry below. There is no ambiguity in this because we will only discuss methods applying to reflectometry in the rest of this work and will not need to mention refractometry again.

2.3 Working Principle of Deflectometry

The basic experimental setup for a deflectometric measurement can be seen in Figure 2.3. It consists of a *screen* which can be any structured light source – most commonly a standard liquid crystal display (LCD) attached to a computer. Other researches have used light projector systems [Pér01] or a heated plate which generates patterns in the far infrared – a wavelength where most materials are highly specular reflective [HK05]. In this work we will stick to the well treated path of using a LCD monitor as structured light source.

The screen displays a pattern or a series of patterns which are reflected in the object under investigation (the *surface*). The pattern is distorted through the curvature of the surface and this distorted pattern can be imaged with a standard photo or video camera. Assuming every beam is only reflected once which holds for well be-

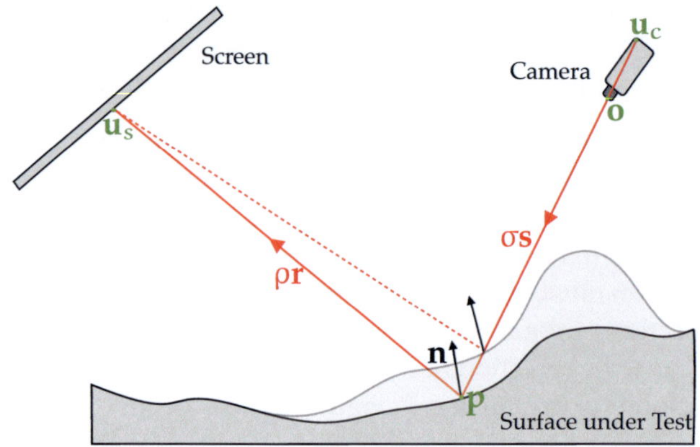

Figure 2.3: Working principle of deflectometry.

haved surfaces and since the object is specular reflective a camera pixel registers either exactly one or no beam originating from the screen.

Following the traditions of computer graphics literature, we model the beam's way reversed to the physical direction: for each camera pixel $\mathbf{u}_c$ we can construct a beam through the pixel and the optical center $\mathbf{o}$ of the camera with a unit length direction vector $\mathbf{s}$. This beam crosses the surface in the point $\mathbf{p} = \mathbf{o} + \sigma\mathbf{s}$. The surface's unit normal vector in $\mathbf{p}$ is $\mathbf{n}$. The direction of the reflected beam can be directly calculated via the reflection law

$$\mathbf{r} = \mathbf{s} - 2(\mathbf{n} \cdot \mathbf{s})\mathbf{n} \qquad (2.1)$$

The reflected beam crosses the screen in the corresponding screen pixel $\mathbf{u}_s = \mathbf{p} + \rho\mathbf{r}$.

We now introduce the simple geometric mapping function (SGMF)

$$l : \mathbb{N}^2 \mapsto \mathbb{R}^2 \qquad (2.2)$$

$$\mathbf{u}_c \rightarrow \mathbf{u}_s \qquad (2.3)$$

which maps the two-dimensional pixel indices $\mathbf{u}_c$ of the camera pixel to the two-dimensional pixel indices of the screen $\mathbf{u}_s$. Note that the screen pixels are assumed to be known subpixel accurate (i.e. not discrete, but continuous values) – we will discuss in chapter 3 how this can be achieved.

Given an absolute calibration of the experimental setup – i.e. camera intrinsics $\mathbf{A}$ and extrinsics $\check{\mathfrak{E}}$ relative to the world coordinate frame and the screen plane patch – i.e. the screens orientation, position and dimensions – $\mathfrak{S}$ in world coordinates – we can augment the SGMF to the geometric mapping function (GMF) which incorporates the calibration data to transfer the pixel indices to coordinates in the world coordinate frame:

$$L : \mathbb{R}^2 \mapsto \mathbb{R}^3 \tag{2.4}$$

$$\mathbf{u}_c \to \mathbf{u}_s. \tag{2.5}$$

Note that we do not make a notational difference between two-dimensional pixel coordinates and their corresponding three-dimensional coordinates in space as this will be clear from context. We will call the tuple $\mathcal{D} := (\mathbf{A}, \check{\mathfrak{E}}, \mathfrak{S}, l)$ a deflectometric measurement. When we work with more than one we will make explicit to which object we refer by using a point, e.g. $\mathcal{D}_1.\check{\mathfrak{E}}$ are the extrinsics of the camera used in the deflectometric measurement 1.

Choosing the optical center $\mathbf{o}$ as origin and using the reflection law (2.1) we can write the GMF as

$$L(\mathbf{u}_c) = \sigma\mathbf{s} + \rho\mathbf{r} \tag{2.6}$$

$$= \sigma\mathbf{s} + \rho(\mathbf{s} - 2(\mathbf{n} \cdot \mathbf{s})\mathbf{n}) \tag{2.7}$$

$$= (\sigma + \rho)\mathbf{s} - 2\rho(\mathbf{n} \cdot \mathbf{s})\mathbf{n}. \tag{2.8}$$

This equation contains two unknowns – namely ρ and σ – and therefore has a one-dimensional solution space. This also becomes apparent in Figure 2.3 when we look at the alternative dotted beam path: it has the same GMF – the same camera view ray sees the same screen pixel – but the surface is crossed at another point and its normal is different. In fact for each σ, one can construct a surface normal so that the GMF does not change. This ambiguity cannot be solved and makes it impossible to reconstruct a surface from just one measurement. We will discuss methods for the reconstruction using more than one measurement in section 4.

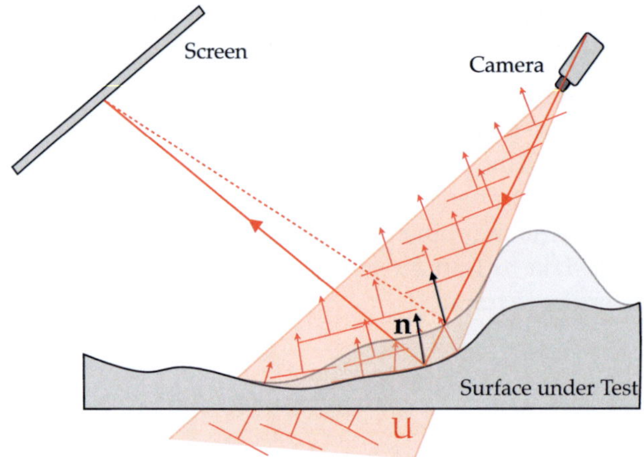

Figure 2.4: The ambiguity interpreted as potential normals in a measurement volume U of the camera.

This ambiguity can also be captured in a different interpretation of a single deflectometric measurement: it can be seen as a normal map in the complete view volume of the camera as depicted in Figure 2.4. That is for each point $\mathbf{p}$ in the measurement volume of the camera U one can find a normal vector $\mathbf{n}$ that does not contradict the measurement. We will see in chapter 4 that this formulation is often much more useful.

We also introduce a different formulation of the reflection law using the mathematical framework from chapter 1. We will show that the reflection is just another dual quaternion, so it fits in nicely with the rest of our mathematical constructions. We start out by defining a dual quaternion that will rotate the view ray $\check{\mathbf{s}} := \mathbf{s} + \epsilon\,\mathbf{s}_m$ – which can be constructed from a pixel and the camera origin – around the line $\check{\mathbf{n}} = \mathbf{n} + \epsilon\,(\mathbf{n} \times \mathbf{p})$ through $\mathbf{p}$ in direction of the normal vector. The resulting line is pointing from the screen to the object, we therefore multiply it by -1 to get the line $\check{\mathbf{r}} = \mathbf{r} + \epsilon\,\mathbf{r}_m$ and therefore the same situation as in Figure 2.3.

According to (1.86), the transforming dual quaternion is simply $(0, \mathbf{n}) + \epsilon\,(0, \mathbf{n} \times \mathbf{p})$, i.e. the augmented line representation for the normal. We now introduce names for the dual and real parts of the dual quaternions we need in our calculations.

$$\mathbf{n}_l = (0, \mathbf{n}) \qquad \mathbf{n}_m = (0, \mathbf{n} \times \mathbf{p}) \tag{2.9}$$

$$\mathbf{s}_l = (0, \mathbf{s}) \qquad \mathbf{s}_m = (0, \mathbf{s}_m) \tag{2.10}$$

$$\mathbf{r}_l = (0, \mathbf{r}) \qquad \mathbf{r}_m = (0, \mathbf{r}_m) \tag{2.11}$$

The reflected line $\check{\mathbf{r}} = \mathbf{r} + \epsilon\, \mathbf{r}_m$ can therefore be found using (1.80)

$$\mathbf{r} + \epsilon\, \mathbf{r}_m = -(\mathbf{n}_l + \epsilon\, \mathbf{n}_m)(\mathbf{s}_l + \epsilon\, \mathbf{s}_m)(\overline{\mathbf{n}}_l + \epsilon\, \overline{\mathbf{n}}_m) \tag{2.12}$$

$$= \mathbf{n}_l \mathbf{s}_l \mathbf{n}_l + \epsilon\, (\mathbf{n}_l \mathbf{s}_m \mathbf{n}_l + \mathbf{n}_l \mathbf{s}_l \mathbf{n}_m + \mathbf{n}_m \mathbf{s}_l \mathbf{n}_l) \tag{2.13}$$

$$= (0, \mathbf{s} - 2(\mathbf{n} \cdot \mathbf{s})\mathbf{n}) + \epsilon \tag{2.14}$$

$$\left(0, \mathbf{s}_m + 2((\mathbf{n} \cdot \mathbf{s})(\mathbf{p} \times \mathbf{n}) + (\mathbf{s} \cdot (\mathbf{p} \times \mathbf{n}) - \mathbf{s}_m \cdot \mathbf{n})\mathbf{n})\right) \tag{2.15}$$

This result is derived in section A.3.

The real part of this equation is obviously equivalent to (2.1). The benefit in using this formulation of the reflection law is that it is now only a dual quaternion multiplication. We can therefore concatenate rigid transformations and reflections using only dual quaternion operations.

2.3.1 Sensitivity on Slope Changes

We now discuss the sensitivity of the deflectometry on local slope. Figure 2.5 shows how the first situation (blue) becomes the second situation (green) by rotating the blue surface by the angle α. We are now interested in the amount the value of the SGMF in the corresponding camera pixel changes. Geometrically speaking, we search for $\Delta u_s(\alpha)$. Observing that $\alpha = \gamma_1 - \gamma_2$ and using this to get

$$\beta_2 - \beta_1 = \gamma_1 - (\gamma_2 - \alpha) = 2\alpha \tag{2.16}$$

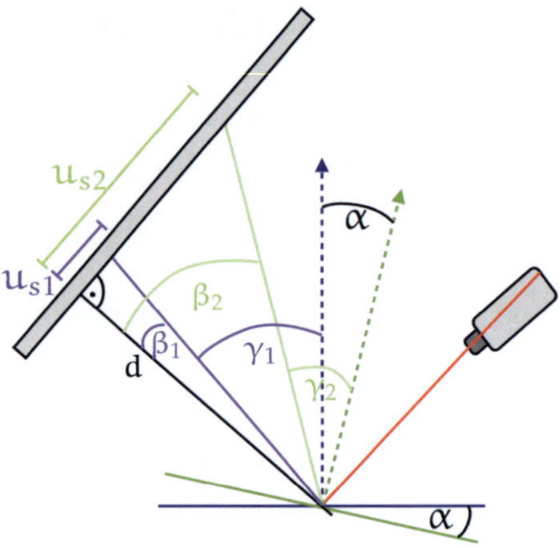

Figure 2.5: Dependency of the measurement on slope changes. The surface under test is rotated around the surface point by an angle of α. This translates the blue situation into the green.

we get

$$\Delta u_s = u_{s2} - u_{s1} \tag{2.17}$$

$$= d(\tan \beta_2 - \tan \beta_1) \tag{2.18}$$

$$= d \tan(\beta_2 - \beta_1)(1 + \tan \beta_1 \tan \beta_2) \tag{2.19}$$

$$= d \tan(2\alpha)\left(1 + \frac{u_{s1} u_{s2}}{d^2}\right). \tag{2.20}$$

In the literature the implicit assumption is usually made that $u_{s1} = 0$. This is an implication that only holds for very few points on the surface and therefore is wrong most of the time. However, it allows certain lower bound approximations to the theoretical limits of the deflectometry which we will discuss in section 2.4, therefore we note that given this approximation, we get

$$\Delta u_s = d \tan 2\alpha. \tag{2.21}$$

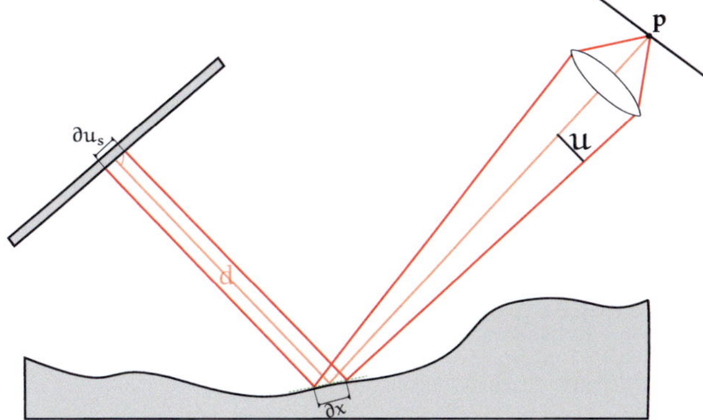

Figure 2.6: Elements involved in the determination of the uncertainty principle of the deflectometry.

2.4 Theoretical Limits

The deflectometry measures the local slope of the surface under test. It has been established in [WH03] that the transfer of local slope allows for three-dimensional reconstruction with less two-dimensional raw data than if local height is transfered as information. This is understandable: noisy height data will yield even noisier data after differentiation – the other way around is noise suppressing. But some information is lost: the slope does not contain global information like the position of the object relative to the camera – the additive constant in the primitive of the slope data. This additive constant is directly related to the ambiguity of the deflectometry. A more elaborate discussion of the deflectometric measurement in the framework of information theory can be found in [Kna06].

The geometrical limits of the deflectometric measurement process have been developed by [Kam05] and [Hor06]. A number of simplifications have been done in both approaches, namely the assumption that the surface can locally be approximated by a parabola and that the light beam arrives close to the axis of this parabola.

Another approximation to the geometric limits of the deflectometric measurement process is the formulation as uncertainty principle – without doubt inspired by the formulation from Heisenberg done in physics. This formulation was originally published in [Häu99] and later again in [Kna06] but only in German so far. We consider it important and interesting, therefore a very similar deduction is provided here in our own words. Figure 2.6 gives an overview of some of the geometrical relationships.

Note that this discussion is rather limited and some of the approximations are quite crude. We will make the following assumption:

1. The camera and screen do not consist of pixels. That is, we ignore spacial discretization artifacts.

2. We will assume that the SGMF is measured using a phase shift technique with a wavelength of U. Therefore $u_s = \frac{\varphi}{2\pi}U$. This is not strictly necessary but allows us to give a formula for the contrast K directly.

3. We assume that the screen is orthogonal to the line connecting screen and surface point. This is the special case discussed in section 2.3.1 – we therefore will only be able to derive a lower bound on the possible precision of the deflectometry. Analogously to (2.21) a change of the surface slope of $\delta\alpha$ will give a change in the phase angle of the corresponding SGMF value of

$$\delta\varphi = \frac{2\pi}{U}d\tan(2\delta\alpha) \approx \frac{2\pi}{U}2d\,\delta\alpha. \qquad (2.22)$$

The cameras aperture size is $2u$, using the approximation for the area seen by the camera pixel on the surface by Rayleigh [Dem] we approximate

$$\delta x = \frac{\lambda}{\sin u}. \qquad (2.23)$$

Herein, λ is the frequency of the light emitted by the area of the screen seen by the camera pixel. For large values of d a good approximation of the screen area is

$$\delta u_s \approx 2d\tan u \approx 2d\sin u = 2d\frac{\lambda}{\delta x}. \qquad (2.24)$$

According to Lampalzer [Lam03], the uncertainty in the phase angle is proportional to the reciprocal contrast $1/K$ for all phase shifting techniques – we reach a similar conclusion in section 3.2.3. Lampalzer defines the quality factor $Q := \frac{1}{K}\frac{2\pi}{\delta\varphi}$ which captures the impact of physical noise on the acquired phase angle.

The contrast can be found by convoluting the area seen by the camera with the displayed sine wave – an approximation for the result is

$$K = \frac{\sin(\pi\,\delta u_s/U)}{\pi\,\delta u_s/U}. \tag{2.25}$$

By substituting $\delta\varphi$ in (2.22) with Q and in turn substituting in K we get:

$$\delta\alpha = \frac{U}{2dQ}\frac{1}{K} \tag{2.26}$$

$$= \frac{\pi}{2dQ}\frac{\delta u_s}{(\sin\pi\delta u_s/U)} \tag{2.27}$$

$$\geqslant \frac{\pi}{2dQ}\delta u_s. \tag{2.28}$$

Now, we use the result (2.24) to replace δu_s which yields the final result

$$\delta\alpha\,\delta x \geqslant \frac{\pi\lambda}{Q}. \tag{2.29}$$

This represents an uncertainty principle for the deflectometry. Due to the many approximations in this derivation and the fact that most of them have a substantial impact on the uncertainty this represents a very rough lower bound estimation which is mainly interesting for its theoretical appeal and the limits it imposes on the deflectometry as a method. There is no deflectometric setup in existence that even comes close to this lower bound.

2.5 Problem Formulation

We now have a good understanding of the basic principle of the deflectometric measurement process and its geometric relationships and limitations. This empowers us to formulate the problem which we will investigate for the rest of this work more strictly.

Prerequisites Given is an experimental setup where camera and screen can be moved between measurements. The object under test is assumed to be at least partially reflective and continuous. It is located anywhere in space and is never moved, altered, or changed in any other way. The camera is a perfect pinhole. The location of screen and camera is assumed to be known up to a sufficient precision in a Cartesian world coordinate frame – we will relax this particular assumption in chapter 5. We also have the means available to acquire a sub pixel precise SGMF for each camera and screen position constellation – how we do this is discussed in chapter 3.

Goal After having chosen a plane as two-dimensional coordinate system with base vectors $\mathbf{e}_x$ and $\mathbf{e}_y$, the goal is to provide for any point $a\mathbf{e}_x + b\mathbf{e}_y$ with $a, b \in \mathbb{R}$ either that no surface information is available, or the distance z the surface has from this point along a vector $\mathbf{t}$ that has a component in the direction $\mathbf{e}_z := \mathbf{e}_x \times \mathbf{e}_y$. For some methods we will directly choose $\mathbf{t} := \mathbf{e}_z$, other methods travel along view rays of the camera with the pixel plane being the coordinate system of choice.

The goal is therefore to provide a two-and-a-half-dimensional representation of the surface in the form of a function

$$S : \mathbb{R}^2 \mapsto \mathbb{R} \tag{2.30}$$

$$a\mathbf{e}_x + b\mathbf{e}_y \to z \tag{2.31}$$

in our Cartesian world coordinate frame.

The variables x and y are fully continuous, i.e. we do not constrain ourself to any discrete grid but we also do not require that S must be representable in an algebraic form. We are therefore not concerned with interpolating measurement data but rather to acquire as precise pointwise measurements as possible.

Later, we will relax the prerequisites to also allow for imprecise extrinsics and screen positions and extend the goal to simultaneously providing information about the surface and improving the knowledge about camera and screen positions.

Chapter 3

Determining the Simple Geometric Mapping Function

The first step in any deflectometric measurement process is to find the simple geometric mapping function $l(\mathbf{u}_c)$ (SGMF). This function is a mapping from the pixel plane of the camera to the pixel plane of the screen and encodes the geometric information of which camera pixel sees which screen pixel. The measurement can be done by displaying a series of images on the screen which uniquely encode the position of the displaying pixel. There are various well known techniques for encoding this information in shapes, a nice overview is given by [SPB04]. However, for the measurements done in deflectometry a sinusoidal phase shift is optimal as we will argue in the section 3.2.1. We will also briefly discuss the gray code, the most important binary encoding scheme and discuss the weaknesses of binary patterns using it as example.

A generalized concept of the SGMF is the optical transfer function (OTF) which is very common in literature of optics [Pér01, Jäh95]. It is the Fourier transformed locale impulse response of the optical system consisting of camera and target. The optical transfer function also encodes information about how the intensity of a screen pixel spreads over neighbouring pixels and how the phase of the light is shifted by the optical system.

The SGMF is essentially the collection of the local maxima of the absolute value of the OTF:

$$l(\mathbf{u}_c) = \operatorname{argmax}_{\mathbf{u}_s} |\text{OTF}(\mathbf{u}_s, \mathbf{u}_c)|. \tag{3.1}$$

Here, $\mathbf{u}_c$ is the camera pixel, $\mathbf{u}_s$ is the screen pixel and argmax will return the argument where the function becomes maximal.

The SGMF makes it possible to investigate an object independent of local reflectivity variations because it is a purely geometric construction. Also, the SGMF is significantly easier to measure than the full OTF. The basic idea in each measurement is to encode the pixel coordinates on the screen in patterns. The camera sees the patterns reflected in the target and can revert the coding to get the information which screen pixel is reflected into which camera pixel.

3.1 Gray code

A simple approach to coding the discrete information of pixel indices is to use a binary pattern. To encode for example 1024 pixel indices, one could use ten pictures and encode the indices as binary pattern – each image would represent one bit. Straight forward base 2 encoding is very susceptible to measurement errors though.

The gray code G was first described in a patent [GRA53] but without a proper name. It found widespread use quickly because of its many desirable properties. First, it has a minimal hamming distance between neighbouring codes of exactly one; that is if only one bit is toggled during the measurement process the error in the result is minimal.

Following, we write code words in binary because this translates well into black and white pixels in n images for the patterns on the screen. The gray code is defined recursively. The four code words needed for encoding $n = 2$ bits of information are

$$G[2] = (00, 01, 11, 10).$$ (3.2)

For higher dimensions, the gray code becomes

$$G[n] = \left(0\|x \middle| \forall x \in G[n-1]\right) \dots \left(1\|R(x) \middle| \forall x \in G[n-1]\right).$$ (3.3)

Here, $\dots$ denotes tuple concatenation and $\|$ defines concatenation of digits:

$$a\|b = a\, 2^n + b.$$ (3.4)

The reflection operation R on a string of n bits $b_0\|b_1 \dots \|b_m$ is defined as

$$R(b_0\|b_1 \dots \|b_m) = b_m\|b_{n-1} \dots \|b_0.$$ (3.5)

The gray code shares the fundamental problems of binary patterns discussed in section 3.2.1 which means that information is lost when watching the screen through an optical system. Like all binary codings, it also has the disadvantage of not providing a sub-pixel precise mapping of screen to camera pixels. Also, to reach a correct mapping, the numbers of code words must be higher than the resolution of the camera; so for a camera with 1600x1200 pixels resolution, one needs at least a screen with equal resolution and $\lceil \log_2 1600 \rceil + \lceil \log_2 1200 \rceil = 22$ images for encoding both pixel index directions.

These arguments make the gray code hardly suitable as only encoding mechanism used in deflectometry. Nevertheless it has its use as a helper to unwrap multi phase shift measurements (see section 3.2.5.1).

3.2 Phase Shifting

3.2.1 The Benefit of Using Sinusoidal Patterns

This work will mostly use sinusoidal patterns as base functions for the measurement of the SGMF. The reason for this becomes apparent

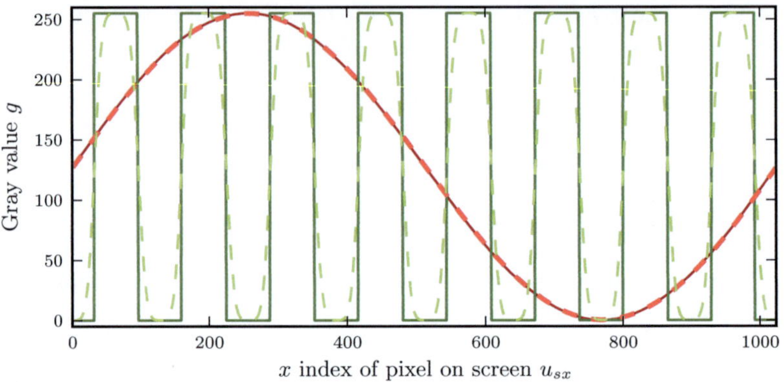

Figure 3.1: Low pass filter on binary and sinusoidal signals. The dashed lines are the solid lines after filtering and normalizing. It can be seen that the sinusoidal shape with only one frequency does not change shape, while the binary signal gets rounded.

when one considers the following problem: The camera optic used in the measurement process can only be focused on one distance. It is beneficial to have the focus point on the surface under inspection to have the best spatial resolution. This implies that the screen's reflection in the surface – and therefore the shape that it displays – will be blurred in the camera picture.

Blurring through defocus in an optical system is equivalent to low pass filtering the image acquired through the optical system. Blurring will always reduce the contrast in the image, i.e. it will reduce the amplitude of all frequencies the image contains – but higher frequencies will be more affected than low frequencies. This low pass characteristic will therefore remove information from the image.

Figure 3.1 shows the effect of low pass filtering of a binary and a sinusoidal screen pattern. As the sinusoidal image only has one frequency, there will not be any reduction in the information it transmits - only its amplitude will be reduced. The binary signal though is rounded. This makes it impossible to decide in some areas if one sees a white or a black pixel of the screen.

3.2.2 Phase Shift Algorithms

Interferometry, Moiré measurement or correlating measurement techniques can often be reformulated or implemented as phase shift measurements. The original idea is presented in [BHG$^+$74]; it has since branched out and is now used in various applications. The specialization that is most suitable for deflectometry has been advanced significantly by Surrel. Especially [Sur96] and [Sur97] are highly educational and give a good overview. Our treaty of the topic is based on [Pér01], the first work to apply these techniques to deflectometry.

In [RFHJ08] we treated a mathematically similar problem and derived the math and error analysis in the context of correlating Time-of-flight camera systems. Measuring the SGMF with phase shift boils down mathematically to very similar formulas. The problem is slightly simpler in this case because no correlation is needed. We will proceed by deriving results for our special case; this is to lay the groundwork for the more involved discussion of statistical and systematical errors in the phase measurement. This error analysis is a novel contribution to the field, as it has not been done before in the context of deflectometry.

Each screen pixel has a unique coordinate $\mathbf{u_s} = (u_{sx}, u_{sy})^\top$ with the origin being the upper left corner of the screen. The goal is to derive the SGMF using the varying intensities of camera pixels while displaying varying patterns on the screen. The problem can easily be decomposed into first deriving u_{sx} and then deriving u_{sy}; therefore the problem becomes one-dimensional and will be discussed as such below. The screen pixel index is now the scalar u_s. U_s defines the number of pixels in the current direction (so for example if $u_s = u_{sx}$ then $U_s = 1024$ for a screen with 1024x768 pixels of resolution).

The screen displays sinusoidal patterns of the form

$$I_s^k(u_s) = A_s \cos(\varphi + \Phi_k) \tag{3.6}$$

with $\varphi := \frac{2\pi}{U_s} u_s$ and Φ_k being well defined phase shifts which will change with each displayed image. Of course, no real display can show negative values – an offset must be added to the screen pattern to shift it into a positive domain. Without loss of generality, we ignore this offset in our discussion. The camera pixel that receives the reflected light from u_s will measure changing intensity values of the form

$$I_c^k = c + A\cos(\varphi + \Phi_k) \tag{3.7}$$

with the following three unknowns: the constant background illumination c, the amplitude A that is recorded by the camera and φ which contains the information about the index u_s of the reflected screen pixel. It is understood that I_c^k, c and A are also dependant on the camera pixels position u_c; this dependency is omitted for clarity as we can solve this problem for each camera pixel individually.

For the following derivations it is useful to write (3.7) as a complex function using the Euler equation. We further set $\Phi_k := 2\pi\frac{k}{N}$ with $k \in [0\ldots N-1]$; this is a special case of the general phase shifting where the shifts between each image are by a fixed angle. The number of images is N. This is the only phase shifting that is used in practice and was already introduced in [BHG$^+$74].

The received intensity then becomes:

$$I_c^k = \frac{A}{2}\left(e^{-2\pi i\frac{k}{N}}e^{-i\varphi}\right) + c + \frac{A}{2}\left(e^{2\pi i\frac{k}{N}}e^{i\varphi}\right) \tag{3.8}$$

Given that $N \geqslant 3$, the solutions for the three unknowns can be derived by using the following formulas:

$$A = \frac{2}{N}\left|\sum_{k=0}^{N-1} I_c^k e^{-2\pi i\frac{k}{N}}\right|, \tag{3.9}$$

$$\varphi = \arg\left(\sum_{k=0}^{N-1} I_c^k e^{-2\pi i\frac{k}{N}}\right), \tag{3.10}$$

$$c = \frac{1}{N}\sum_{k=0}^{N-1} I_c^k. \tag{3.11}$$

We will now show that this solution is optimal in the least squares sense (as long as $N \geqslant 3$, otherwise there is no solution at all). To show this we introduce the abbreviations $z := e^{i\varphi}$ and $y := e^{\frac{2\pi i}{N}}$.

Using these we can write (3.8) in matrix notation:

$$\underbrace{\begin{pmatrix} 1 & 1 & 1 \\ y^1 & \overline{y}^1 & 1 \\ \vdots & \vdots & \vdots \\ y^{N-1} & \overline{y}^{N-1} & 1 \end{pmatrix}}_{:= \mathbf{M}} \cdot \underbrace{\begin{pmatrix} \frac{A}{2}z \\ \frac{A}{2}\overline{z} \\ c \end{pmatrix}}_{:= \mathbf{p}} = \underbrace{\begin{pmatrix} I_c^0 \\ \vdots \\ I_c^{N-1} \end{pmatrix}}_{:= \mathbf{d}} \tag{3.12}$$

To derive the general least squares solution

$$\mathbf{p}_{opt} = (\mathbf{M}^*\mathbf{M})^{-1}\,\mathbf{M}^*\,\mathbf{d} \tag{3.13}$$

we need to calculate the Moore-Penrose inverse of $\mathbf{M}$ which is easy here because $\mathbf{M}$ mainly contains roots of unity:

$$(\mathbf{M}^*\mathbf{M})^{-1}\,\mathbf{M}^* = \frac{1}{N}\mathbf{M}^* \tag{3.14}$$

which yields the least squares solution

$$\mathbf{p}_{opt} = \frac{1}{N}\mathbf{M}^*\,\mathbf{d} = (\frac{A}{2}z, \frac{A}{2}\overline{z}, c)^\mathsf{T}, \tag{3.15}$$

which is equivalent to (3.9)–(3.11) since $A = |Az|$ and $\varphi = \arg(Az)$.

While $N = 3$ will theoretically suffice to find the parameters of the phase shift, it is very susceptible to noise. Therefore, the most important special case is $N = 4$ which yields the best trade-off between accuracy and measurement speed. [ZS95] have shown that higher values for N suppress errors in the second harmonics better, but as discussed in section 3.2.1 we do not have higher harmonics and the increased effort doesn't give better results in our experiments. A similar finding was also reported by [Kam05]. We will therefore concentrate on $N = 4$ in all further discussion and also $N = 4$ was used in all experiments. However we will discuss a small variation to this algorithm using six images below.

By setting $N = 4$, the above results become very simple:

$$\varphi = \arctan\left(\frac{I_c^3 - I_c^1}{I_c^0 - I_c^2}\right), \tag{3.16}$$

$$A = \frac{1}{2}\left|I_c^0 - I_c^2 + i(I_c^3 - I_c^1)\right|, \tag{3.17}$$

$$c = \frac{1}{4}(I_c^0 + I_c^1 + I_c^2 + I_c^3) \tag{3.18}$$

The result (3.16) is also mentioned in [Kam05] and [Pér01], [Kam05] also uses (3.17) to mask out areas where there is no modulation. We will show in the next section that (3.17) also has a direct relationship to the quality of the phase measurement; it is therefore more useful than just as masking value because it carries some confidence about the phase measurement.

3.2.3   Statistical Error Propagation

Naturally, the images acquired from the camera show a statistical noise in the gray values. We assume a normal process with a variance of σ^2 which is not exactly true (see [Jäh95]) but an acceptable approximation. The question is now what impact this noise has on the calculation of A, φ and c. We are more interested in the relations than in the exact numbers as we expect significant errors in the measurements of involved quantities as we will discuss in section 3.2.4. Therefore, we will use Gaussian error propagation and only discuss the case $N = 4$. Again, we can build upon some of our prior work [Rap07] as the math is very similar.

A discussion of the phase noise in relation to the signal to noise ratio of the camera can be found in [Sur97] and more detailed in [Rat95]. In the recent work [FPT11] a quantitative noise model was introduced that represents the phase noise with parameters from the EMVA 1288 standard for camera systems [Jäh10]. All those are more specific models than the one we will introduce but also contain the relationship which we will derive now.

We start out by defining a function f which maps the raw camera data $\mathbf{I_c} := (I_c^0, I_c^1, I_c^2, I_c^3)^\top$ to the data vector $(A, \varphi, c)^\top$:

$$f : \mathbb{R}^4 \mapsto \mathbb{R}^+ \times [0, 2\pi] \times \mathbb{R} \tag{3.19}$$

$$\mathbf{I_c} \to (A, \varphi, c)^\top. \tag{3.20}$$

The calculation of the Jacobian becomes easier when we separate f into two functions. It then becomes

$$f = \chi_2 \circ \chi_1 \tag{3.21}$$

with

$$\chi_1(\mathbf{I_c}) = \begin{pmatrix} \frac{1}{2} & 0 & -\frac{1}{2} & 0 \\ 0 & -\frac{1}{2} & 0 & \frac{1}{2} \\ \frac{1}{4} & \frac{1}{4} & \frac{1}{4} & \frac{1}{4} \end{pmatrix} \mathbf{I_c} \quad \text{and} \quad \chi_2(A, \varphi, c) = (\Omega^{-1}(A, \varphi), c). \tag{3.22}$$

We used the mapping to polar coordinates $\Omega(a, b) = (A\cos(b), A\sin(b))$ here. The calculation of the Jacobian

$$Df(\mathbf{I_c}) = D\chi_2(\chi_1(\mathbf{I_c})) \, D\chi_1(\mathbf{I_c}). \tag{3.23}$$

can now be easily done by using the relationship

$$D\left[\Omega^{-1}(\Omega(a, b))\right] = (D\Omega(a, b))^{-1}. \tag{3.24}$$

It is

$$Df(\mathbf{I_c}) = \frac{1}{2} \begin{pmatrix} \cos\varphi & -\sin\varphi & -\cos\varphi & \sin\varphi \\ -\frac{1}{A}\sin\varphi & -\frac{1}{A}\cos\varphi & \frac{1}{A}\sin\varphi & \frac{1}{A}\cos\varphi \\ \frac{1}{2} & \frac{1}{2} & \frac{1}{2} & \frac{1}{2} \end{pmatrix} \tag{3.25}$$

This can now be used to determine the linear error propagation

$$\text{Var}(A, \varphi, c) = Df(\mathbf{I_c}) \, \text{Var}(\mathbf{I_c}) \, Df(\mathbf{I_c})^\top \tag{3.26}$$

$$= \sigma^2 Df(\mathbf{I_c}) \, Df(\mathbf{I_c})^\top \tag{3.27}$$

$$= \text{diag}(\frac{1}{2}, \frac{1}{2A^2}, \frac{1}{4})\sigma^2. \tag{3.28}$$

The most interesting result is that the variance of the phase (and with it the variance of the pixel position u_s) is inverse proportional to the amplitude squared:

$$\text{Var}(u_s) \propto \frac{1}{A^2}. \tag{3.29}$$

The amplitude can therefore be used as a confidence gauge for the pixel position: if the amplitude is high so will be the confidence of having measured the pixel position with high precision.

3.2.4 Systematical Errors

There are two kinds of systematic measurement errors that are considered in this work. The first one are small errors in the amount Φ_k each phase gets shifted and the second is the non linear response curve of the screen used in the experiments. Those effects will be discussed consecutively now.

3.2.4.1 Phase Shifting Error

Due to the nature of our display device, there is a quantization effect on every continuous function we want to display. This quantization error can be represented by replacing Φ_k in the formulas via $(1 + \eta)\Phi_k$ with a small unknown $\eta \in \mathbb{R}$. The question we ask now is how this error affects our calculated data.

[KLSS88] was the first to investigate systematical shift errors. [ZS95] showed that the error for the four sample algorithm is of order $O(\eta\varphi)$ in the phase calculation. They propose a variation to the four sample phase shifting algorithm: they use $N = 4$ in (3.6) but take two more pictures with $k = 4, 5$. Those images seem redundant, but make the phase shifting error η observable. This leads to only a constant offset $O(\eta)$ in the phase measurement which is non crucial because it can be compensated by virtually shifting the displaying screen. The algorithm reads as

$$\varphi = \arctan\left(2\frac{I_c^3 + I_c^4 - I_c^1 - I_c^2}{I_c^0 + I_c^1 + I_c^4 + I_c^5 - 2(I_c^2 + I_c^3)}\right) - \frac{\pi}{4}. \tag{3.30}$$

Note that [ZS95] investigated the formula for more complex patterns that can be represented as a Fourier series of degree two:

$$I_c^k = c + \sum_{m=0}^{2} A_m \cos(\varphi_m + m\Phi_k). \tag{3.31}$$

This has no relevance for our application as our signal will not contain higher frequencies in practice.

3.2.4.2 Non Linear Response of Screen

The upper plot in Figure 3.2 shows the response of a standard LCD monitor. The desired gray values $g = [0 \dots 255]$ are linear, the response is not. In fact, the response curve is very well approximated by a parabola without constant terms:

$$M(g) = p_1 g + p_2 g^2 \tag{3.32}$$

Note that p_1 and p_2 will depend strongly on the viewing angle and weakly on the position of the pixel on the screen. Also, the response for other LCD technologies will look different [FPT10]. An optimal fit in the least squares sense and the corresponding residual plot are also depicted in the Figure.

What happens to the phase shifting algorithms when we assume a response curve with unknown p_1 and p_2? For each phase shift position k, the camera pixel will record an intensity image of

$$\tilde{I}_c^k := c + M(I_s^k) = c + M(A \cos(\varphi + \Phi_k)). \tag{3.33}$$

We assume a linear relationship between luminosity on the screen and recorded pixel gray value in the camera here. As before, we dropped dependency on the camera pixel $\mathbf{u}_c$ for clarity; the dependency on the screen pixel is implicit in φ.

For the four sample algorithm, the final formula for intensity (3.18)

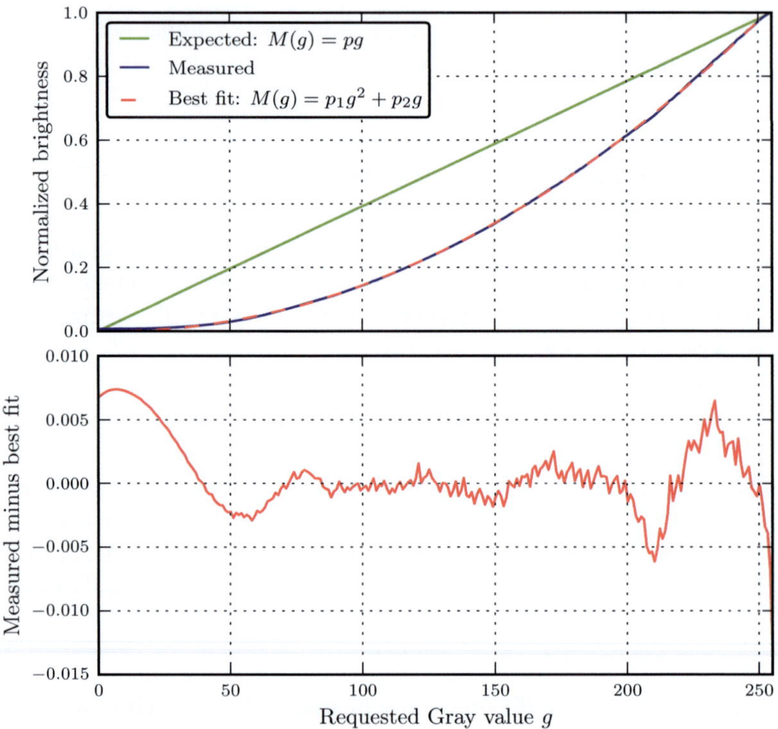

Figure 3.2: Plot of the behaviour of a real LCD monitor. The measured intensity values have been normalized. The bottom plot shows the error between the measurements and the best fit curve of the upper plot.

then becomes

$$\tilde{c} = \frac{1}{4}(\tilde{I}_c^0 + \tilde{I}_c^1 + \tilde{I}_c^2 + \tilde{I}_c^3) \tag{3.34}$$

$$= \frac{1}{4}c + \frac{1}{4}\sum_{k=0}^{3}\left(p_2 A^2 \cos^2(\varphi + \Phi_k) + p_1 A(\cos\varphi + \Phi_k)\right) \tag{3.35}$$

$$= c + \frac{1}{4}2p_2 A^2 \tag{3.36}$$

$$= c + \frac{1}{2}A^2 p_2. \tag{3.37}$$

The measured intensity is therefore a significant overestimation depending on the square of the physical amplitude displayed in the corresponding display pixel. The calculation for the amplitude reads

$$\tilde{A} = \frac{1}{2}|\tilde{I}_c^0 - \tilde{I}_c^2 + i(\tilde{I}_c^3 - \tilde{I}_c^1)| \tag{3.38}$$

$$= \frac{1}{2}|2Ap_1 \cos\varphi + i(2Ap_1 \sin\varphi)| \tag{3.39}$$

$$= Ap_1 \tag{3.40}$$

So we measure a quantity that depends linearly on the correct amplitude. We will first do an auxiliary calculation for the phase:

$$\tilde{I}_c^3 - \tilde{I}_c^1 = A^2 p_2^2 \left(\cos^2(\varphi + \frac{3}{2}\pi) - \cos^2(\varphi + \frac{1}{2}\pi)\right) \tag{3.41}$$

$$+ Ap_1 \left(\cos(\varphi + \frac{3}{2}\pi) - \cos(\varphi + \frac{1}{2}\pi)\right) \tag{3.42}$$

$$= A^2 p_2^2 \left(\sin^2\varphi - (-\sin\varphi)^2\right) + Ap_1 (\sin\varphi + \sin\varphi) \tag{3.43}$$

$$= 2Ap_1 \sin\varphi \tag{3.44}$$

With the similar result for $\tilde{I}_c^0 - \tilde{I}_c^2$, we get for the phase calculation

$$\tilde{\varphi} = \arctan\left(\frac{\tilde{I}_c^3 - \tilde{I}_c^1}{\tilde{I}_c^0 - \tilde{I}_c^2}\right) \tag{3.45}$$

$$= \arctan\left(\frac{\sin\varphi}{\cos\varphi}\right) \tag{3.46}$$

$$= \varphi. \tag{3.47}$$

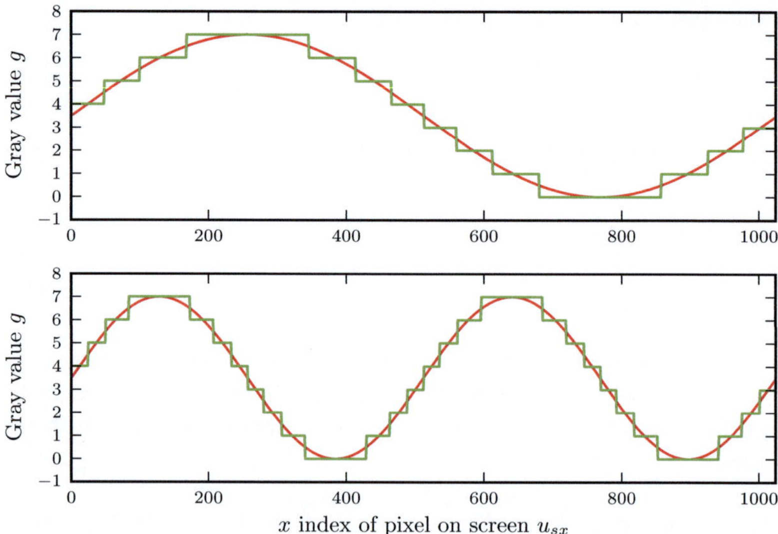

Figure 3.3: Real world problems with phase shifting algorithms. The top plot shows errors introduced by discretization. The bottom plot shows an improvement concerning the discretization by displaying two periods of the sinusoidal pattern. However, the phase is no longer isomorph to the pixel index x.

The most important property – the phase calculation – is passed through correctly. But if the non linearity of the screen is not corrected, we will get incorrect information for the amplitude and the intensity.

The derivation for the six sample algorithm is comparable to the four sample algorithm and also yields a similar result: intensity and amplitude are changed, but the phase is correctly estimated.

3.2.5 Phase Unwrapping Algorithms

To find an unambiguous solution for the phase shift algorithm, only a single sine wave must be displayed on the screen. This leads to practical problems though: The limited contrast of real screens and the discretization on the screen as well as in the camera lead to a reduction of the spatial resolution when only one frequency is used. The problem is depicted in Figure 3.3. The top plot shows in green what a display with only 8 shades of gray would display if the red curve was requested and the screen is linear in its response. It is clear that for example all pixels between 190 and 380 have the same phase. The Figure below shows the same plot, but now two periods of the wave are displayed on the screen at the same time. The areas of ambiguity are significantly smaller than in the top plot, but the phase is no longer isomorph to the pixel index x: even in the red curve, the indices 150 and 662 have the same phase value.

Phase unwrapping algorithms are attempts to increase local contrast by displaying more than one period on the screen while keeping the phase and the pixel index isomorph to each other. Since extra measurements are needed to get an absolute and unique phase unwrapping we call the methods discussed below *temporal unwrapping alogrithms* compared to *spatial unwrapping algorithms* which do not need extra data but can only provide a relative phase unwrapping which is not useful for three-dimensional reconstruction [ZLY07, ZZB09, SS03]. Consequently, we will not explore spatial unwrapping algorithms in this work. The next few sections will discuss various temporal unwrapping algorithms.

3.2.5.1 Combining Phase Measurements with Gray Code

A very simple approach is to combine a single phase shift measurement with a high local contrast and ambiguity and a binary encoding scheme like the gray code. The data from the gray code can be used to resolve the ambiguity in the phase shift measurement. Such a system was originally described in [SCR99], recent advancements were done by [ZSXS11]. It has found a wide application in fringe pattern reconstruction. Binary patterns work best when the camera is focused on the screen which is not desirable for deflectometry. This leads to reconstruction errors where the binary pattern is unsharp.

Figure 3.4 shows an unwrapping done using this scheme. The final result is smooth except for noisy jumps. These are the areas where

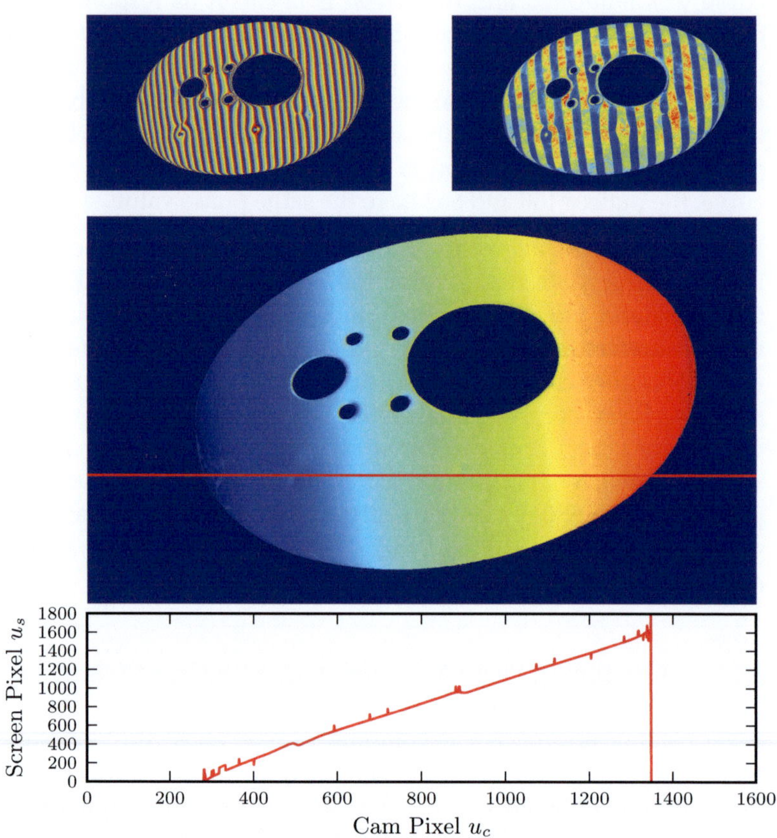

Figure 3.4: Unwrapping a horizontal phase shift using gray code. The small images show one image from the phase shift measurement and one from the gray code measurement. The big picture is the final result, the plot below shows the cross section marked in red in the image. Clearly visible are the jumps in the unwrapping at the phase shift borders.

the phase shift measurement wraps but the absolute phase measurement from the gray code predicts the wrong phase number. These errors could be corrected using a postprocessing step given the surface is known to be smooth. This experiment needs eleven images for the gray code plus four images for the phase shift measurement.

3.2.5.2 Chinese Remainder Wave Lengths

Another approach to unwrapping is supported through the Chinese remainder theorem which we will state and proof in a simplified version.

Chinese Remainder Theorem. *Given are the following simultaneous congruences:*

$$s \equiv s_1 \quad \mathrm{mod}\ p_1 \tag{3.48}$$

$$s \equiv s_2 \quad \mathrm{mod}\ p_2 \tag{3.49}$$

with $\{s, s_1, s_2\} \in \mathbb{R}$ and $\{p_1, p_2\} \in \mathbb{N}$ being coprime to each other, i.e. $GCD(p_1, p_2) = 1$. There now exists a unique solution $\{a_1, a_2\} \in \mathbb{Z}$ for the equation

$$1 = a_1 p_1 + a_2 p_2. \tag{3.50}$$

It then follows

$$s = (s_1 a_2 p_2 + s_2 a_1 p_1) \quad \mathrm{mod}\ (p_1 p_2) \tag{3.51}$$

Proof. Equation (3.50) is a special case of Bézout's identity. Multiplying (3.48) with p_2, (3.49) with p_1 and (3.50) with s yields the three equations

$$sp_2 \equiv p_2 s_1 \quad \mathrm{mod}\ (p_1 p_2) \tag{3.52}$$

$$sp_1 \equiv p_1 s_2 \quad \mathrm{mod}\ (p_1 p_2) \tag{3.53}$$

$$s = s a_1 p_1 + s a_2 p_2 \tag{3.54}$$

Substituting the first two equations in the third directly yields the final result. $\qquad\square$

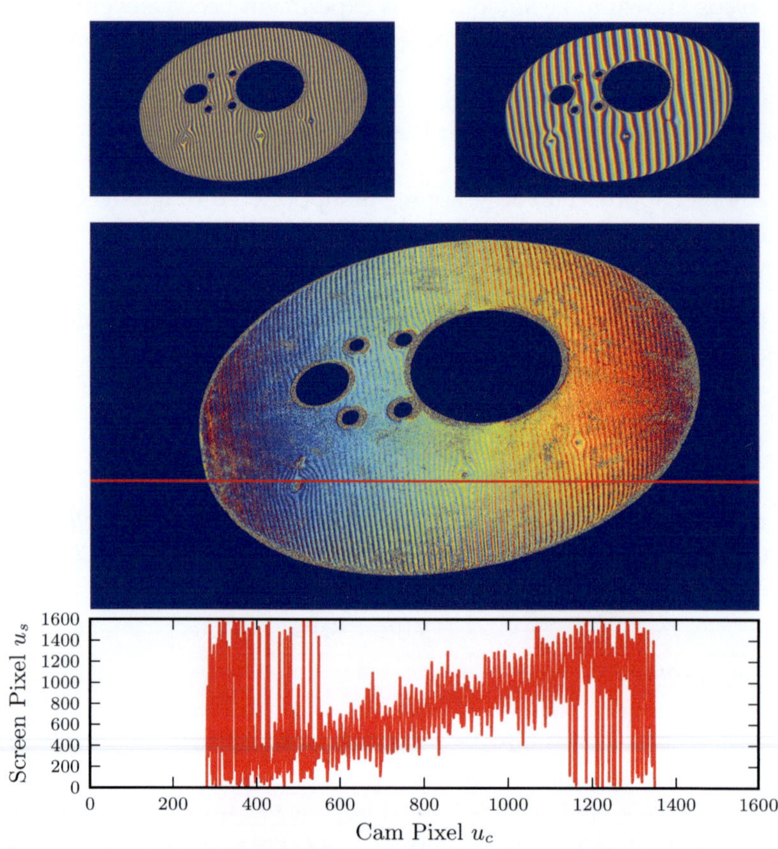

Figure 3.5: Unwrapping a horizontal phase shift using the Chinese remainder theorem . The small images show one image from each phase shift measurement. The big picture is the final result, the plot below shows the cross section marked in red in the image. The reconstruction is extremely noisy due to the modulo arithmetic involved.

A more involved proof and a generalization for simultaneous congruences with more than two elements can be found in [Yan02].

A practical solution for real world phase shift measurements would be now to choose two wave lengths $\{\lambda_1, \lambda_2\}$ such that $\lambda = \lambda_1\lambda_2$ is bigger than the maximum pixel index we need to encode. It is useful to choose one wavelength small to have a high local contrast. For our screen with a resolution of 1600 pixels one can only use $\lambda_1 = 5^2 = 25$ pixels and $\lambda_2 = 2^6 = 64$ pixels as this is the only co-prime factorization. Each pixel will then have two measured values Φ_1 and Φ_2. The Bézout pairs can be found efficiently using the extended Euclidean algorithm [Knu68]. The unique index can then be reconstructed via equation (3.51).

Figure 3.5 shows the result of this unwrapping. The result is terribly noisy which is an effect of the modulo arithmetic involved. For example, with the given lambdas the measurement of $\Phi_1 = 19$ and $\Phi_2 = 73$ would lead to a $\Phi = 969$. Disturbing Φ_1 to 19.1 results in $\Phi = 1026.6$. Therefore, while this approach has a great theoretical appeal and only needs 8 images, its sensitivity to noise makes it unusable when images are acquired with standard industrial video cameras like in our setup.

3.2.5.3 Multi-Phase Shift Measurement

The basic idea of the Multi-Phase Shift (MPS) technique – which is sometimes also called iterative multi wavelength approach – is to incrementally refine the result until no further improvement is measurable. That is, one starts with a basic wavelength of λ_0 that will result in a non ambiguous measurement with a low local contrast and therefore high noise. Now the frequency is doubled, so $\lambda_1 = \lambda_0/2$. This increases the local contrast but will result in ambiguity - the first measurement can now be used to resolve these ambiguities and receive a new result with higher local contrast and no ambiguity. The frequency is now doubled again and the ambiguity resolved as before etc. etc.

This can be stated as an algorithm as follows:

$\Phi_{-1} \leftarrow \Phi_0$

$$k = 2$$

repeat

$\quad \lambda = \lambda_0/k$

$\quad \Phi \leftarrow$ result of phase measurement with λ.

$\quad$**for** each pixel index u **do**

$$n \leftarrow \lfloor \mathrm{abs}\left(\frac{1/k\Phi(u) - \Phi_{-1}(u)}{2\pi}\right)\rfloor$$

$$\Phi(u) = \Phi(u) + n\pi$$

$\quad$**end for**

$\quad \Phi_{-1} = \Phi$

$\quad k \leftarrow 2k$

until $\sum \mathrm{abs}(\Phi - \Phi_{-1}) < \delta$

This technique has the advantage of high precision which is bought with a high number of pictures one needs to acquire. On the other hand the amount of precision can be determined beforehand and the ending criterion δ be chosen accordingly.

This algorithm is well established: it came up in the nineties [HS93] and stays a focus of current research till today [Zha09]. A sample unwrapping can be seen in Figure 3.6. We acquired a multi phase shift measurement with nine wave lenghts, so a total of 36 images. The final and intermediate results are shown below the image. From the bottommost plot it becomes clear that nine wavelengths are not needed – 5 to 6 usually suffice for a good reconstruction. More wave lenghts will increase accuracy at the cost of longer measurement time.

One problem that must be kept in mind when choosing the smallest wavelength for the phaseshift are Moiré patterns. When the wavelength is chosen too small to be properly sampled with the spacial resolution of the camera, Moiré patterns might be seen in the camera instead of the proper wavelengths. This will in fact not improve the unwrapping but make it worse. The minimal wavelength depends on the surface under test and its geometry and must be found empirically before a MPS measurement can be acquired.

The MPS method provides the highest precision which was a priority for this work. We therefore decided to settle on using this method over the others.

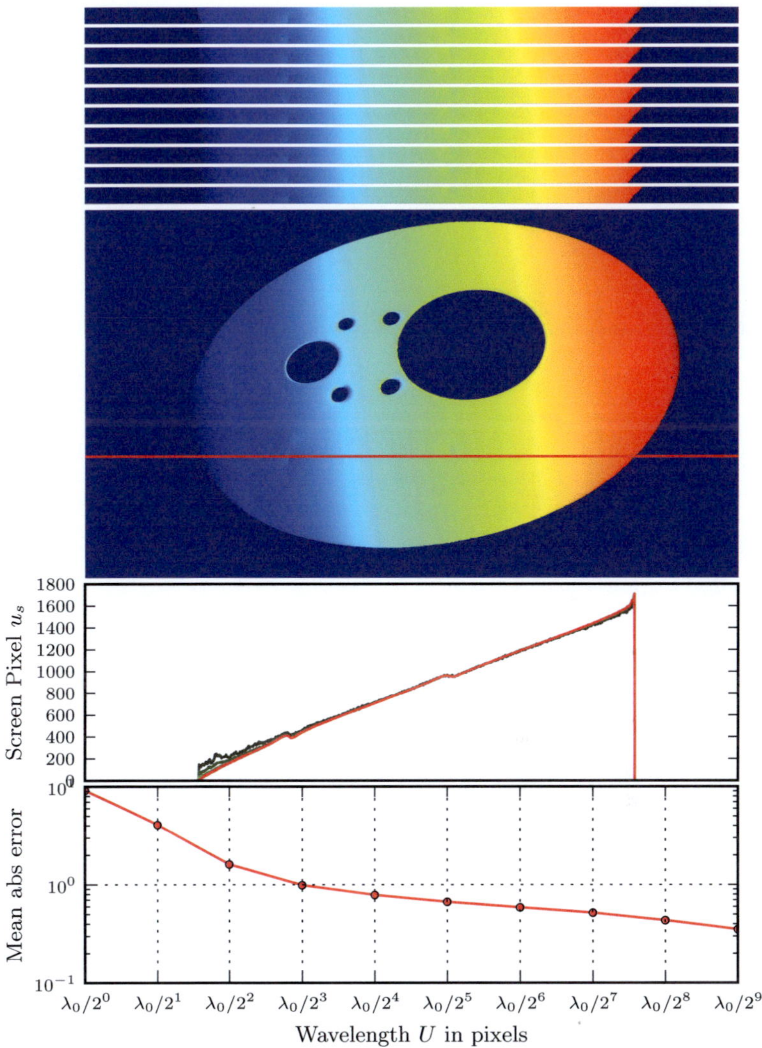

Figure 3.6: Unwrapping a horizontal phase shift using the multi phase shift. The small stripes are the intermediate results around the red line in the image below. The big picture is the final result. The plot below shows the values of the red line in the intermediate results. The final result is red, the intermediate results are in green increasing in brightness. The bottom plot shows the mean absolute error between the final and the intermediate results.

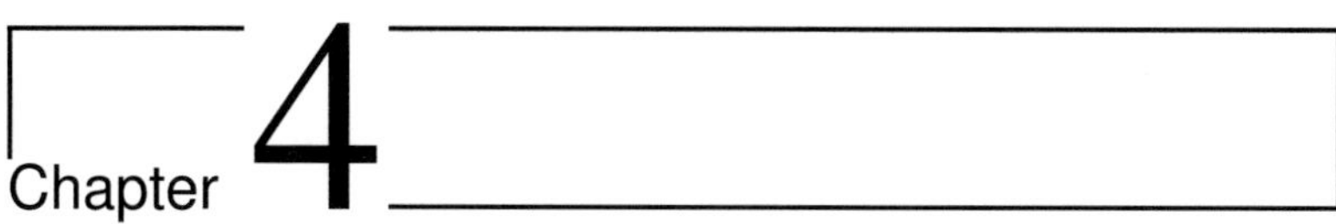

Chapter 4

Surface Reconstruction

When the SGMF is determined up to a sufficient precision and the extrinsic calibration between camera and screen is known, three-dimensional reconstruction can start. This is usually done in two steps: the first is to resolve the ambiguity of a single deflectometric measurement by taking additional data into account. This results in an approximate position of the surface in the form of points in space and the more precise normal vectors of the surface in these points. Acquiring this information will be our concern in this chapter. We will discuss methods that can create arbitrary many points and normals of the surface. We can therefore generate an infinite amount of data describing the surface which means effectively reconstructing it.

Usually, a second and final step is done after this reconstruction: here, one takes the low precision points and the higher precision normal values and interpolates them to get a smooth and continuous representation of the surface. This problem – called height from normals – is also needed for other three-dimensional measurement techniques like the stripe projection. We will not discuss it in this work and instead point to the exhaustive and ever growing literature. For the deflectometry, the traditional method for interpolating the normal data is a Hermite interpolation scheme with radial basis functions [EKH07]. A modern and flexible approach to the problem using a kernel based projection into a higher-dimensional space can be found in [NWT10]. There are also many time proven methods with various fortes and weaknesses like the Frankot-Chellappa algorithm [FC88] or differentiating the data and solving the equivalent Poisson equation [PFT$^+$86].

Every deflectometric measurement can be created by an infinite number of surfaces but the discriminating information to identify the true surface can be encoded in one parameter [Bal08]. This information is inherently not contained in a single measurement and therefore additional data needs to be acquired to find this parameter and resolve the ambiguity.

There have been a lot of suggestions for combining deflectometric information with other techniques: shape-from-shading [BWB06], stereo [KH05], and the degree of polarisation of the reflected light [Hor07] have been proposed. Also combinations with other reconstruction techniques like depth-from-focus or stripe projections can work in some cases. These techniques have in common that the surface needs to fulfill additional constraints – besides being reflective – like having a certain amount of diffuse reflection, having markers attached, or being made of metal. We only want reflectivity as constraint for this work, so this discussion will concentrate on techniques that combine deflectometric measurements to resolve the ambiguity.

We will start our discussion with per-pixel ambiguity solving algorithms. They have the advantage of being completely local, i.e. there is no drift error and no dependence on a starting pixel. They are also trivial to parallelize which is a necessity to reach viable running times of the reconstruction. We will outline our approach to parallelization which drastically reduces running times and makes it possible to combine various methods and generalize them to more than two measurements.

We will continue our discussion with a novel approach to reconstruct the surface using a region-growing integration combined with a consistency evaluation using a second deflectometric measurement. As any numeric integration scheme this will also have a global error drift and depend heavily on the pixel chosen as initial seed.

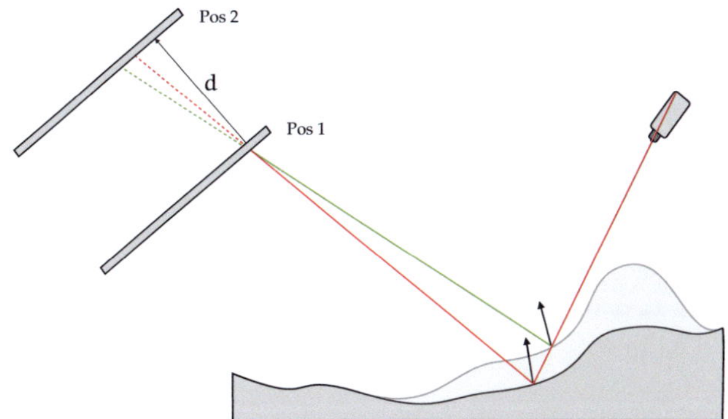

Figure 4.1: Active reflection grating photogrammetry

4.1 Active Reflection Grating Photogrammetry

A simple method to resolve the ambiguity is to use two measurements where only the screen has been moved. This technique is known as active reflection grating photogrammetry (ARGM) [PR01] – sometimes also called regularisation through movement of the screen and is historically the first method proposed to create three-dimensional data from two deflectometric measurements.

The basic idea can be seen in Figure 4.1. The first measurement returns the camera view ray and a point on the first screen. This information has an infinite number of possible beam travel paths – a second possibility is shown in green. When the screen is moved a second screen point becomes known with the next measurement. This point will uniquely identify the correct beam and the point of intersection with the surface can now easily be found via triangulation.

This simple technique has several disadvantages in practical applications. First, the movement of the screen between the two measurements must be precisely known to guarantee a sufficient precision for the intersection points. This usually means that the screen needs to be mounted on a very precise actuator - usually a one-dimensional linear unit. This limits the view volume of the experimental setup considerably. The second problem is that it is difficult to move the screen in such a way that all areas of interest of the surface still reflect it. Especially for convex objects finding two acceptable screen positions is very often a challenge. For this reason and because our experimental setup is not able to move the screen as precise as needed for this technique, we do not further investigate the ARGM in this work.

4.2 Passive Reflection Grating Photogrammetry

An improvement over the ARGM is the passive reflection grating photogrammetry (PRGM) [PT04]. It allows for both screen and camera to be moved which makes it much more usable for real world experiments. However, it does not provide a solution for the calibration problem on its own – that is, all screen and camera positions must be known prior to reconstruction.

The basic principle is depicted in Figure 4.2. The algorithm proceeds as follows: first, any point $\mathbf{p}$ on the view ray of the first camera is chosen. This defines a normal $\mathbf{n}$ which is then tested for validity with the second measurement. This is done as follows: The point $\mathbf{p}$ is projected into the camera plane of the second measurement yielding a view ray. This ray is reflected using $\mathbf{n}$ and intersected with the screen plane of the second measurement. This results in a screen pixel $\mathbf{u}'_s$. Its distance $\Delta\mathbf{u}_s$ from the truly seen screen pixel $\mathbf{u}_s$ given through the SGMF is the value we try to minimize by varying the point $\mathbf{p}$. We call $\Delta\mathbf{u}$ the inconsistency of the reconstruction.

We will now state the implementation of this algorithm as well as its pseudocode. The algorithm takes the following arguments: A search volume V which must contain the complete surface, a step parameter $t \in \mathbb{R}$ and two deflectometric measurements $\mathcal{D}_i$ consisting of

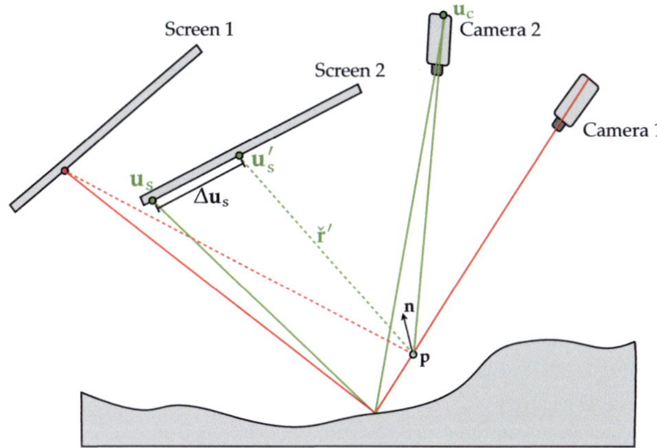

Figure 4.2: Passive reflection grating photogrammetry

camera extrinsics represented by a dual quaternion $\mathcal{D}_i.\check{\mathfrak{E}}$, camera intrinsics represented by a homogeneous matrix $\mathcal{D}_i.\mathbf{A}$ the screen plane $\mathcal{D}_i.\mathfrak{S}$ and a SGMF $\mathcal{D}_i.l$. The algorithm then proceeds independently for each camera pixel $\mathbf{u}_c$ of the first camera. The corresponding view ray through V is sampled into a list of points $P = [\mathbf{p}_1 \ldots \mathbf{p}_m]$. We then proceed to finding the point with the smallest inconsistency $\Delta\mathbf{u}_s$. We use its two neighbouring points in P to initiate a ternary search [Knu68] to find the non-discrete minima $\mathbf{p}_{min}$ on the ray. We then output $\mathbf{p}_{min}$ and its corresponding inconsistency $\Delta\mathbf{u}_s$ for this pixel. Since this algorithm can be run independently for each pixel and because we need to sample the SGMF of the second measurement sub pixel precise – i.e. we need a two-dimensional interpolation of the SGMF – the algorithm is very well suited for running on a graphics processing unit (GPU). The GPU provides a high number of processors which can work simultaneously on different pixels. This provides high scalability and maximal parallelism. Also, all GPUs provide hardware accelerated interpolation in images because it is frequently needed in video games. This has a dramatic impact on the running time of our software: the implementation on the GPU is up to 10.000 times faster than running the same algorithm on the CPU. Our GPU implementation is done in OpenCL [M$^+$09] – the bridging between Python on the CPU and OpenCL on the GPU is facilitated by the excellent PyOpenCL library [KPL$^+$12].

The pseudocode reads as follows.

Require: $\mathcal{D}_1, \mathcal{D}_2, V, t$

    function EVALUATE_POINT$(\mathbf{p}, \mathcal{D}_1, \mathcal{D}_2)$
        $\mathbf{n} \leftarrow \text{normal\_at}(\mathbf{p}, \mathcal{D}_1)$
        $\mathbf{u}_c \leftarrow$ RESECT_INTO_CAMERA$(\mathbf{p}, \mathcal{D}_2)$ ▷ $\mathbf{u}_c$ that sees this point
        $\check{\mathbf{s}} \leftarrow$ VIEW_RAY$(\mathbf{u}_c, \mathcal{D}_2)$
        $\check{\mathbf{r}}' \leftarrow$ REFLECT$(\check{\mathbf{s}}, \mathbf{p}, \mathbf{n})$ ▷ Reflect $\check{\mathbf{s}}$ in $\mathbf{p}$ around $\mathbf{n}$
        $\mathbf{u}'_s \leftarrow$ PLANE_LINE_MEET$(\check{\mathbf{r}}, \mathcal{D}_2.S)$
        $\mathbf{u}_s \leftarrow \mathcal{D}_2.l(\mathbf{u}_c)$
        return $\|\mathbf{u}'_s - \mathbf{u}_s\|$
    end function

    $\text{rv} \leftarrow []$
    for $\mathbf{u}_c$ in all camera pixels of cam 1 **do**
        $\check{\mathbf{s}} \leftarrow$ VIEW_RAY$(\mathbf{u}_c, \mathcal{D}_1)$
        $\mathbf{p} \leftarrow$ optical center of cam 1
        $\mathbf{p}_{\min} \leftarrow \mathbf{p}$
        $\min = \infty$
        while $\mathbf{p} \in V$ **do**
            $\text{val} =$ EVALUATE_POINT$(\mathbf{p}, \mathcal{D}_1, \mathcal{D}_2)$
            if $\text{val} < \min$ **then**
                $\min \leftarrow \text{val}$
                $\mathbf{p}_{\min} \leftarrow \mathbf{p}$
            end if
            $\mathbf{p} \leftarrow \mathbf{p} + t \cdot \text{Real}(\check{\mathbf{s}})$
        end while
        $\mathbf{p}_{-1} \leftarrow \mathbf{p}_{\min} - t \cdot \text{Real}(\check{\mathbf{s}})$
        $\mathbf{p}_{+1} \leftarrow \mathbf{p}_{\min} + t \cdot \text{Real}(\check{\mathbf{s}})$
        $\min, \mathbf{p}_{\min} \leftarrow$ TERNARY_SEARCH$(\mathbf{p}_{-1}, \mathbf{p}_{+1}, \text{Real}(\check{\mathbf{s}}), \mathcal{D}_1, \mathcal{D}_2)$
        $\text{rv.APPEND}(\mathbf{p}_{\min}, \min)$
    end for
    return rv

A fundamental problem with this algorithm is that there are surfaces
that can have many local minima with a similar value. It is possi-

ble that the algorithm returns the wrong minimum then. In fact, it has been shown that one can always find two surfaces given two measurements which cannot be distinguished using this technique [WB11]. Also, this technique reports many false minima around the edges of the view volume which usually can be filtered out by only considering points where min is smaller than a chosen value δ. This happens when the overlapping volume for one view ray is very small and only far away from the surface. A minimal value will exist – sometimes even with a small $\mathbf{u}_s$ – and therefore a point will be reported though no physical surface point exists at this position. The PRGM algorithm also suffers from the hole effect, which we will discuss in the context of the normal comparison algorithm in section 4.3.2.

4.2.1 Generalization to more Measurements

If more than two deflectometric measurements are available which provide data for the same volume, the algorithm can be augmented to consider them all. Given M measurements, we can replace EVALUATE_POINT in the algorithm by a new function COMPLETE_EVALUATE that reads as follows:

> **function** COMPLETE_EVALUATE($\mathbf{p}, \mathcal{D}_1, \ldots, \mathcal{D}_M$)
>   **return** $\sum_{i=2}^{M}$EVALUATE_POINT$(\mathbf{p}, \mathcal{D}_1, \mathcal{D}_i)/(M-1)$
> **end function**

Since the original evaluation function is not symmetric in the arguments $\mathcal{D}_1$ and $\mathcal{D}_2$ the first measurement remains the one defining the view rays along which the search volume is traversed. A direct result of this non-symmetry is that the generalization increases the running time only linearly with the number of measurements. Similarly – but harder to quantify – the improvement in the quality of reconstruction is expected to only improve 'linearly' since only a linear amount of new data is considered as well. If the first measurement $\mathcal{D}_1$ is somehow flawed the whole reconstruction will be flawed as its information is defining the reconstruction.

While this kind of augmentation of the algorithm is only possible through the much improved running times of a GPU implementation, this generalization is not straightforward to implement: To make use of the image interpolation hardware on the GPU, the SGMF are passed as textures to the calculating OpenCL kernel. The number of textures passed to a kernel cannot be varied and must be known before compiling the bytecode for the GPU. But with M measurements, we also need to pass M textures to the kernel. We solved this constraint using meta programming: As soon as M is known, the OpenCL source code for the right number of measurements is generated from a template text. It is then compiled and executed on the GPU from our calling program running on the CPU. The source code is cached, so that the time overhead of creating and compiling the GPU kernel needs only to be done once for a given number of measurements and image sizes. A nice benefit of this technique is that we can also benchmark the GPU before compiling the source code and therefore choose ideal compile and call parameters for the specific hardware and host computer configuration.

4.3   Comparing Normal Predictions

An approach similar to the PRGM is the comparison of the normal predictions (NC).

The basic principle is visualized in Figure 4.3. The two measurements used in this technique are interpreted as normal fields in their search volumes V_i. The technique proceeds by choosing a two-dimensional grid – visualized as the blue dotted lines – and traversing the overlapping volume $V = V_1 \cap V_2$ along a search direction $\mathbf{d}$ while minimizing the difference in the predictions.

Of course, some kind of metric needs to be introduced to compare two normals:

$$m : \mathbb{R}^3 \times \mathbb{R}^3 \mapsto \mathbb{R}^+ \tag{4.1}$$

$$m(\mathbf{n}_1, \mathbf{n}_2) \to c \tag{4.2}$$

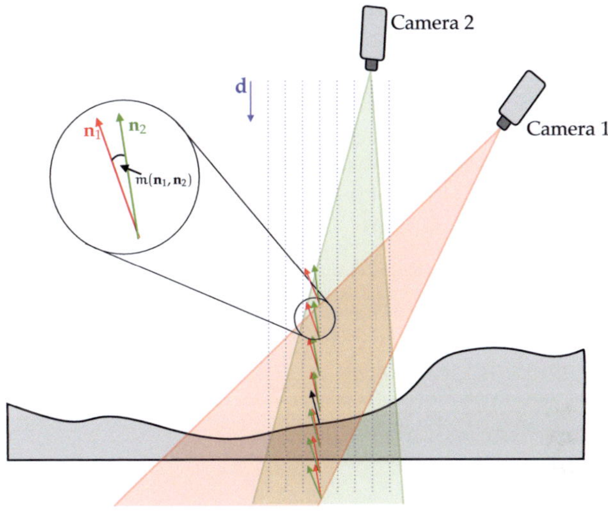

Figure 4.3: Comparing Normal predictions

The two obvious choices are

$$m_1(\mathbf{n}_1, \mathbf{n}_2) = 1 - |\mathbf{n}_1 \mathbf{n}_2| \qquad (4.3)$$

$$m_2(\mathbf{n}_1, \mathbf{n}_2) = \|\mathbf{n}_1 \times \mathbf{n}_2\| \qquad (4.4)$$

which both work equally well. The first one was preferred in the implementation because it is slightly faster to compute. As for the PRGM, we call the value $m(\mathbf{n}_1, \mathbf{n}_2)$ the inconsistency of the current $\mathbf{p}$.

Like with the PRGM, there is no guarantee for a single global minimum along any chosen search direction given any surface. Simple line search algorithms are therefore not sufficient here and might converge on a local minima only. We implemented the NC in a similar fashion to the PRGM using two steps in the minimization: First, the inconsistency values are sampled on the search line in the complete search volume. Two points bracketing the point with the smallest value $\mathbf{p}_{\min}$ are then used as starting point for a refinement procedure. As before, a ternary search using the points $\mathbf{p}_{min} - \mathbf{d}$ and $\mathbf{p}_{min} + \mathbf{d}$ works well.

The algorithm therefore needs as input a grid $G = \{(x_1, y_1, z_1)\ldots\}$ which is chosen to have all starting points laying on a plane, a search volume U that is usually cubic, a metric m to compare two normals, a search direction $\mathbf{d}$, and a step size $t \in \mathbb{R}$. The complete algorithm can then be stated as follows:

Require: $\mathcal{D}_1, \mathcal{D}_2, U, G, z_{start}, \mathbf{d}$

```
function EVALUATE_POINT(p, D₁, D₂)
    n₁ ← normal_at(p, D₁)
    n₂ ← normal_at(p, D₂)
    return m(n₁, n₂)
end function


rv ← []
for each p ∈ G do
    pmin ← p
    min = ∞
    while p ∈ V do
        val = EVALUATE_POINT(p, D₁, D₂)
        if val < min then
            min ← val
            pmin ← p
        end if
        p ← p + t · d
    end while
    p₋₁ ← pmin − t · d
    p₊₁ ← pmin + t · d
    min, pmin ← TERNARY_SEARCH(p₋₁, p₊₁, d, D₁, D₂)
    rv.APPEND(pmin, min)
end for
return rv
```

The algorithm is very similar to the PRGM – the biggest difference is the EVALUATE_POINT function. Therefore it is not surprising that its implementation shares the same ideas: it is most efficiently done on a graphics processing unit because it can be executed in parallel for many pixels at once and because it needs interpolation in both SGMF images. The algorithm has some advantages over the PRGM though: it is more flexible because the search direction and the density of the grid are all parameters that can be adjusted to the need of the user and the EVALUATE_POINT function is significantly cheaper to compute.

4.3.1 Generalization to more measurements

We now generalize the normal comparison algorithm to $M \geqslant 2$ measurements. Similarly as we have done for the PRGM we will introduce a new function called COMPLETE_EVALUATE that can handle more measurements.

> **function** COMPLETE_EVALUATE$(\mathbf{p}, \mathcal{D}_1, \ldots, \mathcal{D}_M)$
>   **return** $\frac{2}{(M-1)M} \sum_{i<j}$EVALUATE_POINT$(\mathbf{p}, \mathcal{D}_i, \mathcal{D}_j)$
> **end function**

We use the fact that the metric m used in the EVALUATE_POINT function is symmetric and therefore the whole function is symmetric. This gives us the opportunity to compare each distinct pair of measurements with each other. There are $\frac{(M-1)M}{2}$ distinct pairs for M measurements which explains the factor before the sum. Averaging is not strictly needed but will keep the metric value in the same range even if we can only use a subset of measurements for a certain point in space. The running time of this algorithm increases linearly with the number of distinct pairs – but we also expect that the quality of the reconstruction increases significantly. Also, we do not have the downside of the PRGM that one arbitrary measurement is special for the algorithm – the normal comparison is completely symmetric.

Similar to how we implemented the generalized solution for the PRGM, we also resorted to meta programming for this implementation. All comments made for the PRGM apply here as well.

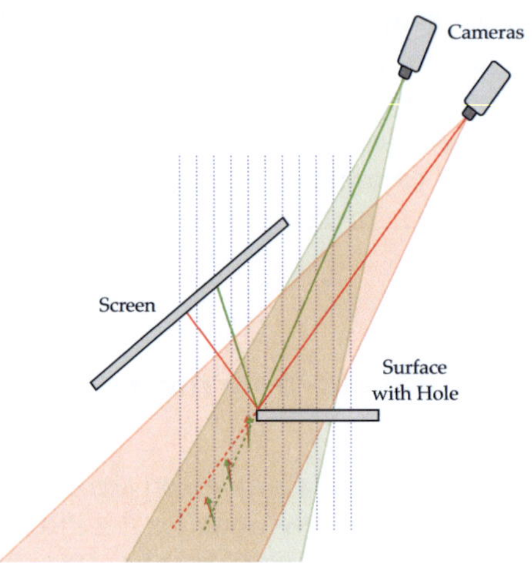

Figure 4.4: Visualization of the hole problem. To the left of the surface and below it, there are points where normal information is available from both measurements – because the resectioning beam crosses the surface – leading to points with minimal inconsistency value though there is no surface there.

4.3.2 False Points on a Plane along the Edges of Holes

Edges and holes in surfaces reveal a fundamental problem in the reconstruction using NC and SGMF. The problem is easiest understood with the NC but it also applies to the PRGM. An example is depicted in Figure 4.4. The reconstruction on the surface works perfectly here, but the search grid continues to the left of the surface. At the sample points where normals are drawn there are indeed normal predictions from both measurements available and therefore also a minimum in the inconsistency. This leads to wrong points around holes. Most of the times, the inconsistency values for these points are big – but not always: the closer to the true surface, the smaller the inconsistency value. This leads to wrong point predictions close to the surface boundary which are not easy to filter out.

The points seem to lie on a line or surface that is related to the view

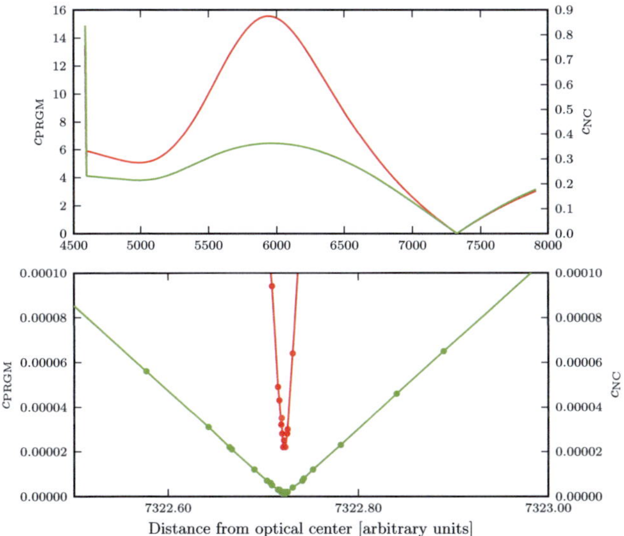

Figure 4.5: Inconsistency values of NC (green) and PRGM (red) along one view ray in a simulated measurement. The top plot shows the complete range, the bottom plot is zoomed on the true minimal value. It also shows the evaluations done by the ternary search.

direction of the cameras. Consequently, using more measurements with different view directions and combining NC with PRGM (see section 4.4) will improve the result but will usually not get rid off all wrong points, especially not close to the true surface.

4.4 Combining PRGM and NC

While the PRGM cannot be easily adapted to work inside the framework of the NC, the other way follows naturally. The grid G for the NC can be chosen to be the pixel plane of the first camera and similarly the search direction $\mathbf{d}$ can be chosen differently for each camera pixel – also to be along the view ray corresponding to the camera pixel. We then traverse the search volume in the same fashion as the PRGM.

This allows us to combine the PRGM and the NC. We use the PRGM algorithm as framework and augment the EVALUATE_POINT function to also include the content of the corresponding NC function. The remaining problem is that the two evaluations return different units as inconsistency values: NC returns a value $c_{NC} \in [0, 1]$ while the PRGM returns a distance on the pixel plane $c_{PRGM} \in \mathbb{R}^+$. Figure 4.5 shows the two inconsistency values on a representative view ray in a simulated measurement. The inconsistency for the PRGM usually has a stronger dynamic while the inconsistency for the NC has usually a higher symmetry around the global minima which is a desirable property for the ternary search. Sometimes – though not in this example – the two functions will have other local minima in different places and only agree at the correct place.

We therefore want to choose a weighting scheme that is dominated by the NC around the minima to retain the symmetry but avoids local minima of only one function. A good compromise is $20c_{NC} + c_{PRGM}$. The 20 is somewhat arbitrary and scales the NC value to be of comparable scale to the PRGM value. When both values are low, the NC value dominates the inconsistency which helps with the symmetry around the minima.

Of course, this algorithm can be generalized to more measurements in exactly the same manner as the PRGM. Once again, the GPU implementation allows us significantly more freedom in improving the algorithm: Considering NC and PRGM and three measurements increases the running time by roughly a factor of ten compared to two measurements and only the NC. This means still reasonable running times of a few seconds with the GPU, but would take hours or days on a CPU.

4.5 Consistency Reconstruction

Theoretically, a surface can be reconstructed from a single deflecto-metric measurement. Given one point on the surface, the deflecto-metric measurement delivers a normal prediction at this point which in turn can be used to find neighbouring points on the surface. The

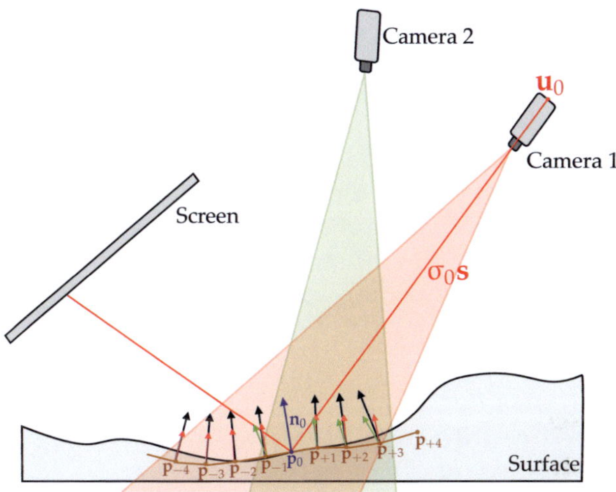

Figure 4.6: Sketch of the consistency reconstruction algorithm and the surface integration by region growing.

general principle is depicted in Figure 4.6. This first order numerical integration scheme will lead to an ever increasing drifting error – however, it is often proposed in the context of deflectometry [Hor07, TLGS05]. Moreover, the initial point on the surface is usually unknown. It can be found by various means – for example via stereo or stripe projection if the surface also has non reflective parts. We find this hearkening back to other techniques unsatisfying. Therefore, we now propose a method that uses a second deflectometric measurement and verifies the consistency of the integrated surface. We call this consistency reconstruction (CR). The algorithm takes two deflectometric measurements $\mathcal{D}_1$ and $\mathcal{D}_2$, an initial camera pixel in the first camera $\mathbf{u}_0$ and an initial distance estimation σ_0. The seeds for the region-growing algorithm – i.e. $\mathbf{p}_0$ and $\mathbf{n}_0$ – can be directly calculated from $\mathbf{u}_0$ and the initial distance σ_0. We proceed by constructing a plane in $\mathbf{p}_0$ with normal $\mathbf{n}_0$ and intersecting it with the view rays of the neighbouring pixels. This results in $\mathbf{p}_{-1}$ and $\mathbf{p}_{+1}$. Repeating the procedure with these points will grow the surface until we reach the end of the view volume V_1 of the camera.

In the overlapping volume $V = V_1 \cap V_2$, we can also get normal predictions from the second measurement. After choosing a metric for vector comparison m and calculating the metric values for all points in V, we can sum up the values. We interpret this value as a residual, i.e. an information about how well the integrated surface matches the prediction of the second measurement and therefore a benchmark for the initial guess σ_0. Let's summarize this integration step:

Require: $\mathcal{D}_1, \mathcal{D}_2, \mathbf{u}_0, \sigma_0$

function FIND_INITIAL_POINT($\mathcal{D}_1, \mathbf{u}_0, \sigma_0$)
    $\check{s} \leftarrow$ VIEW_RAY($\mathbf{u}_0, \mathcal{D}_1$)
    $\mathbf{p}_0 \leftarrow$ optical center of cam 1 $+ \sigma_0 \cdot$ Real ($\check{s}$)
    $\mathbf{n}_0 \leftarrow$ NORMAL_AT($\mathbf{p}_0, \mathcal{D}_1$)
    return $\mathbf{p}_0, \mathbf{n}_0$
end function

function INTEGRATE($\mathcal{D}_1, \mathcal{D}_2, \mathbf{u}_0, \sigma_0$)
    pts $\leftarrow$ 2D_ARRAY_OF_VECTORS_LIKE_PIXELPLANE($\mathcal{D}_1$)
    ns $\leftarrow$ 2D_ARRAY_OF_VECTORS_LIKE_PIXELPLANE($\mathcal{D}_1$)
    residual $\leftarrow 0$
    $\mathbf{p}_0, \mathbf{n}_0 \leftarrow$ FIND_INITIAL_POINT($\mathcal{D}_1, \mathbf{u}_0, \sigma_0$)
    pts[$\mathbf{u}_0$] $= \mathbf{p}_0$
    ns[$\mathbf{u}_0$] $= \mathbf{n}_0$
    done $\leftarrow \{\mathbf{u}_0\}$
    next $\leftarrow \{\mathbf{u}_0\}$
    while next $\neq \{\}$ **do**
        $\mathbf{u} \leftarrow$ next.POP()
        plane $\leftarrow$ PLANE_FROM_POINT_AND_NORMAL(pts[$\mathbf{u}$], ns[$\mathbf{u}$])
        for each pixel neighbour $\mathbf{u}_n \notin$ done of $\mathbf{u}$ **do**
            $\check{s} \leftarrow$ VIEW_RAY($\mathbf{u}_n, \mathcal{D}_1$)
            $\mathbf{p}_n \leftarrow$ PLANE_LINE_MEET($\check{s}$, plane)
            $\mathbf{n}_1 \leftarrow$ NORMAL_AT($\mathbf{p}_n, \mathcal{D}_1$)
            $\mathbf{n}_2 \leftarrow$ NORMAL_AT($\mathbf{p}_n, \mathcal{D}_2$)
            done.INSERT($\mathbf{u}_n$)

```
            next.INSERT(u_n)
            pts[u_n] = p_n
            ns[u_n] = n_1
            residual ← residual + m(n_1, n_2)
        end for
    end while
    return pts, residual
  end function
```

Equipped with this region growing integration, we can now vary σ_0 while minimizing the residual. This will give us a best estimate for an initial distance σ_0 and therefore a unique reconstruction without a known starting value.

Note that this reconstruction will always yield a smooth surface because we are integrating a smooth normal field. The operation is also noise suppressing – any integration is. The operation is also not symmetric in the deflectometric measurements. The first measurement has a stronger impact on the result than the second measurement. One example are holes: When using two measurements where the first one has data for the whole surface while the second one has a hole in its data – i.e. a region where the screen was not reflected due to improper positioning – the resulting reconstruction will either have the hole or not, depending which measurement was taken for integration and which for consistency evaluation.

Of course, the method can easily be extended to use more than two measurements: we can simply take the others as additional consistency tests for the surface we integrate from the first measurement.

This algorithm has the fundamental flaw that even with a perfect start value, the integration error accumulates the farther we get away from the start value. This is to be expected, since this integration is just a first order Euler scheme – the simplest numerical integration scheme there is. But even with more sophisticated numerical integration methods like Runge-Kutta this drift cannot be avoided. The error could only be reduced by running the algorithm with different starting pixels and taking a weighted sum of the results.

4.6 Discussion

The CR has very little appeal: the region growing approach is dependent on neighbouring pixels and therefore not easy to parallelize. Also, the ever increasing integration error the further away one gets from the starting point is a given and cannot be avoided. On the other side, the resulting surface is always smooth and consistent with the normal measurement of the SGMF. The CR is included in this discussion because region-growing integration is so often seen in the literature of deflectometry - hopefully we can help phasing it out over time.

From all algorithms presented here, the NC is the most flexible. It provides freedom in the search direction and the grid positioning. It is also completely symmetric in the measurements and therefore allows for a non linear growth of data when more than two measurements are used. En plus, it also has the cheapest point evaluation function of the presented algorithms. Its inconsistency values are bounded – always between zero and one – which can be a plus or a minus for the algorithm but makes it harder to filter out outliers.

A close runner up is the PRGM. Its point evaluation function is much more expensive than the NC's and it does not scale as nicely when more than two measurements are given due to its non symmetry. It does not need to interpolate in the SGMF of the first camera though and its inconsistency values are unbounded which makes it easier to filter out points with huge values.

Combining NC and PRGM seems a promising approach: one can avoid most of the wrong local minima of both techniques and still keep the nice symmetry of the inconsistency values around the correct minimum from the NC. The price is the loss of the symmetry in the deflectometric measurements and the loss of freedom of choice in the grid and the search direction compared to the NC. Still, the approach will increase robustness of the reconstruction with only a minor cost in the complexity of the algorithm.

We will compare the performance of the algorithms on simulated and experimental data in chapter 6.2.1.

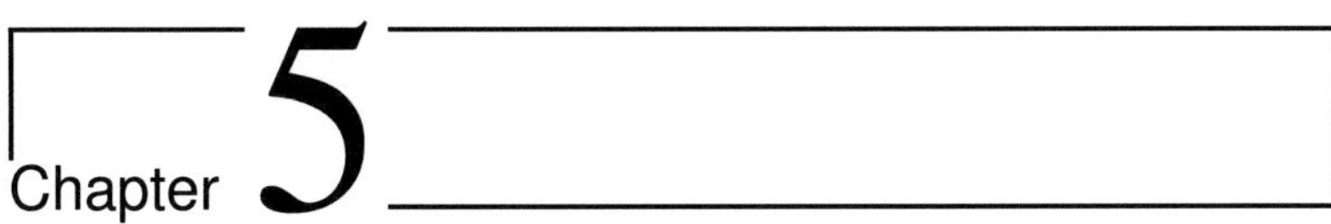

Chapter **5**

Auto-Calibrating Deflectometry

We now know how to reconstruct surfaces given perfect camera and screen extrinsics. Finding these extrinsics is done by calibrating the experimental setup. Static deflectometric setups are usually calibrated using reference objects like flat mirrors [Wer11, HWB10] or with the help of other three-dimensional measurement techniques [Hor07].

But every movement of screen or camera requires a new calibration. And the techniques used for calibration put some constraints on the setup. For example, when using a flat mirror as calibration object, one needs to position screen and camera to acquire a good deflectometric measurement for the mirror. When the mirror is then replaced by another – maybe convex – surface, these positions might no longer be ideal. But the screen and the camera can only be moved together now, or the calibration data is vain.

More flexible approaches usually come at the cost of less precise calibration. For example, in our experimental setup, the extrinsics of the screen can only be deduced by using a camera image of the screen. This method is very imprecise and not good enough for a proper

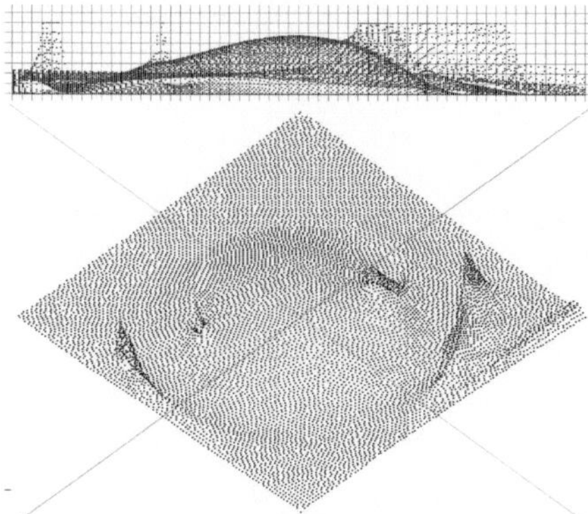

Figure 5.1: NC$(\mathcal{D}_0, \mathcal{D}_2)$ results on Hubbel where the second camera's view direction was wrong by $2°$

deflectometric reconstruction. The camera extrinsics are taken from the position information of an industrial robot arm. This information is quite good, but nowhere near the precision of the calibration one wants to have for accurate deflectometry. We will discuss methods to improve such an approximate calibration. This is not specific to our setup, because an approximate calibration is easy to acquire, for example by attaching markers to screen and camera and using a stereo system observing the deflectometric setup. The methods can also be used to improve the manual calibration of static deflectometric setups.

This chapter describes new methods to improve a rough calibration that only needs to be valid up to a few millimeters. We also simultaneously reconstruct the surface. We believe these methods to be helpful even with static experimental setups but to be indispensable for a dynamic setup like ours. We build this framework on an abstract description of the idea published earlier [RS11].

5.1 Basic Principle

The basic idea is simple: the NC, the PRGM and the NC+PRGM reconstructors all use an inconsistency value to define how well two measurements agree if a point is part of the surface or not. The premise now is that with perfect calibration, this inconsistency value can always become zero on any given view ray that sees a reflected screen pixel. But if the calibration is not perfect, not all inconsistencies will vanish. Figure 5.1 shows the reconstruction result on the simulation object Hubbel (see section 6.1.4) with slightly wrong extrinsics: the camera of the second measurement was off by 2°. This has a dramatic impact on the reconstruction result and also on the inconsistencies. The substantial change in the reconstruction result also hints at the sensitivity of the deflectometry – a fact we use to our advantage in the calibration.

We arbitrarily choose the camera of the first measurement to have correct extrinsics, i.e. it will define our new world coordinate frame. We now proceed as follows: we reconstruct a surface with one of the methods, filter out outliers and take the mean of all inconsistency values in the reconstructed points. We will apply small rigid transformations to the first screen, second screen and second camera and reconstruct again till we converge on a minimal value for the mean of the inconsistencies. That is, we have 18 degrees of freedom to optimize. We could take more measurements into account, but for each new measurement we increase the degrees of freedom by 6 for the screen and 6 for the camera which adds to the optimization running time exponentially. For our experiments, we therefore always only used two measurements.

After convergence, we have not only achieved better calibration data, but we also have a better reconstruction of the surface. We therefore use the surface to improve the calibration and vice versa. This is why we call this method auto-calibrating deflectometry.

The following methods can be used for auto-calibration.

5.1.1 NC Auto-Calibration

The NC algorithm has some nice properties for the auto-calibration. First, the number of points on the reconstruction grid can be freely chosen. If one is only interested in the calibration parameters, the number of points can be reduced to speed up the calibration process. It is also the cheapest algorithm to compute which also affects the runtime of the optimization positively.

The inconsistency values are all between zero and one. This makes it hard to distinguish outlier points from inlier points with a high inconsistency because the current extrinsics are still bad. This often results in the translations of the extrinsics becoming huge in the optimization: if one camera is far from the other, the measurement volumes no longer overlap that much and therefore there are no or few points which often results in a low mean inconsistency. This happens so frequently that the NC on its own is next to useless for auto-calibration.

5.1.2 PRGM Auto-Calibration

The inconsistency values of the PRGM are unbound. It is therefore easier to filter out the completely wrong points which can have inconsistency of 1000 pixels or higher. When we start at a reasonably close approximation of the true extrinsics and only use the points which have an inconsistency of e.g. 20 or lower, we can be quite confident that the points we consider are really part of the surface. We now slightly alter the extrinisics towards a smaller mean inconsistency for these points. In the next reconstruction step which hopefully brought us closer to the true extrinsics we might have more points below 20 that are really part of the surface. The wrongly included points will get a higher inconsistency and eventually drop out of the set of points.

The PRGM is more costly than the NC though. The view rays through the search volume are implicitly defined, therefore the runtime can only be lowered by changing the step length t. In our experiments, the runtime of the PRGM was still sufficiently low to make it a suitable candidate for auto-calibration.

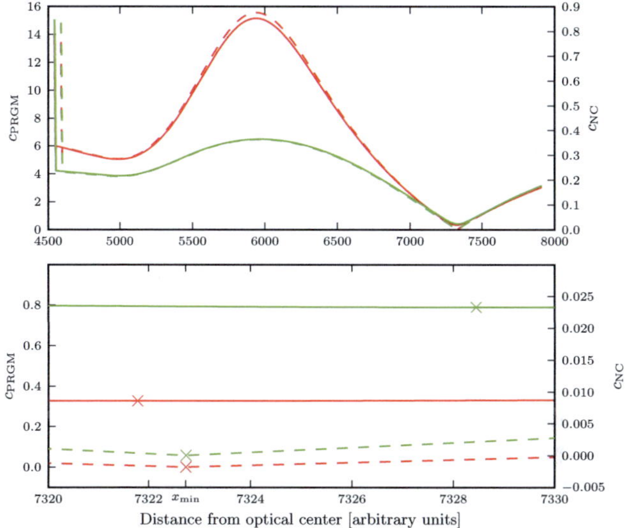

Figure 5.2: Inconsistency values of NC (green) and PRGM (red) along one view ray in a simulated measurement. The top plot shows the complete range, the bottom plot is zoomed on the true minimal value. The dashed lines are using the correct extrinsics, the solid lines are from slightly perturbed extrinsics.

5.1.3 Combi Auto-Calibration

The Combi reconstructor has the same qualities as the PRGM for auto-calibration: Its inconsistency values are unbound and outliers can therefore be removed reliably. But it has an even higher computational cost. However, it has one more benefit: the NC and the PRGM minima are only at the same point if the calibration is good. The minima will not align for wrong calibration data. This should increase the steepness of the descend towards the ideal parameters. Figure 5.2 shows the values for a single view ray from the example in Figure 5.1. In the bottom part of the plot it can be seen that NC and PRGM agree on the true minimum when there is no disturbance but disagree with the wrong extrinsics. This can be used to find the correct parameters.

The impact of auto-calibration will be discussed together with the reconstruction algorithms in chapter 6.

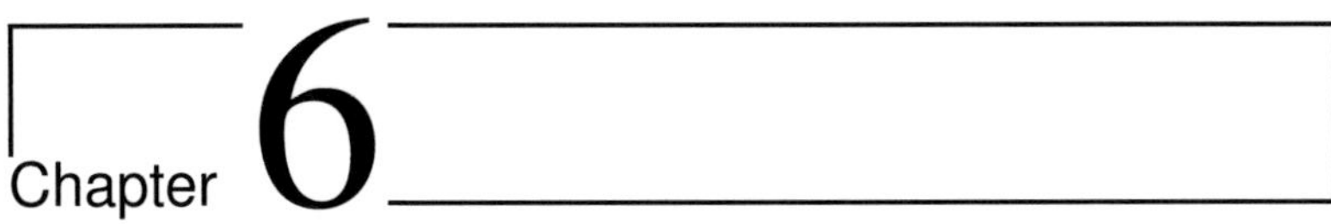

Chapter **6**

Experimental Evaluation

This chapter presents the experiments and simulations performed to test and compare the reconstruction methods explained in chapter 4 and the auto-calibration from chapter 5. We will start by describing our simulation environment and then explain which objects were simulated and how the algorithms performed on them. We will then continue by describing the experimental setup and its static calibration. We will conclude this chapter with some experiments and the performance of auto-calibration and reconstruction on them.

6.1   Simulations

We used different simulated objects and modelled the deflectometric measurement process using a raytracer program. We will start out by presenting our simulation environment, then we will proceed to the simulations and the results of the following methods on them: Normal comparison (NC), Passive Reflection Grating Photogrammetry (PRGM), Combined Reconstruction using NC and PRGM (Combi) and the Consistency Reconstruction (CR). We will also describe the performance of auto-calibration.

After the algorithms were run on the data, the resulting points were filtered using the per point inconsistencies where possible (NC, PRGM, Combi) to weed out outlier pixels. The resulting points were used to compare the algorithms.

6.1.1 Simulation Environment

The tool of choice for the simulation and visualization environment is Blender[1], a free and open-source (FOSS) three-dimensional modelling and rendering tool. It is made with artistic purposes in mind – the rendering engine therefore makes some incorrect simplifications of the physics of light propagation – but for our cases it is well suited because the reflection law is modelled correctly.

Blender has two advantages which makes it the best choice for our purposes: first, it offers complete scriptability which means that we can extract camera extrinsics and screen position from a modelled scene and that we can simulate a phase-shifting screen. The second feature is that it also offers the ability to use custom made texture plugins. This allowed us to model a screen that can be queried for its intensities sub-pixel precise. This gives us a perfect simulation environment to explore our algorithms, their correctness and their performance.

6.1.2 Plane

The simplest case considered here is the reconstruction of a quadratic plane patch laying in the x-y plane at $z = 0$ with a side length of 6. The plane patch S therefore has the following equation

$$S = \{z = 0 \; \forall \; \mathbf{x} = (x, y, z)^{\mathsf{T}} | (-3 \leqslant x \leqslant 3) \wedge (-3 \leqslant y \leqslant 3)\}. \quad (6.1)$$

Three measurements were simulated with three different camera and screen positions – we will designate them as measurements $\mathcal{D}_0$, $\mathcal{D}_1$ and $\mathcal{D}_2$. No measurement shows holes, i.e. in every camera position, every visible pixel of the plane patch reflects a screen pixel

---

[1]`http://blender.org`

and the complete object is visible in all measurements. The measurement $\mathcal{D}_0$ is special because it has the screen middle point and the camera's optical center at $x = y = 0$ and the pixel plane of the camera, the screen and the plane patch are all parallel to each other.

The methods were now tested as follows: each method was tested using very conservative parameters (i.e. small step sizes) to achieve the best possible result. Points with a high inconsistency or with x or y values out of bounds were filtered out. We then fitted a plane through the remaining points. The resulting normal vector $\mathbf{n} = (n_x, n_y, n_z)^\mathsf{T}$ and the distance from the origin d was compared to the correct values of $\mathbf{n} = (0, 0, 1)^\mathsf{T}$ and $d = 0$. We also looked at the mean distance of the points from the correct surface Δ_{mean}. The complete results are compiled in table B.1.

All methods give very good results given the huge measurement volume. There are no outliers left after filtering and the reconstructed points are all very close to the plane. Numerically, the NC method performs consistently best throughout the comparison – its final error is limited by the numerical precision of the simulation – while the CR performs up to two orders of magnitude worse when looking at the mean distance from the correct surface Δ_{mean}. Notably, the NC+PRGM method performs comparable to PRGM alone and does not profit from the slightly better results of the NC method. The NC method gets worse when using all three measurements compared to only considering measurement 1 and 2. The reason for this is that measurement 0 is the nosiest of the three and as it has the same weight in the final result it will increase the final error.

All methods deliver acceptable results in this test and most results are only limited by the precision of the rendering engine and the numerical accuracy of the calculations. The algorithms are therefore all working and suitable for reconstruction, but this simple example does not test the robustness of the algorithms.

6.1.3 Sphere

For the following reconstruction experiments, we acquired data from a sphere with radius 0.5 m with the given reconstruction techniques. The evaluation of the reconstruction was done by fitting a sphere to all points using the RANSAC algorithm [FB81] and comparing the final radius and position to the ground truth. As before, we used three screen-camera position pairs, but due to the convexity of the sphere, only a small part of it could be measured. This shows a big problem with a flat screen: its dimensions must be enormous to acquire data from a strongly convex object. We also filtered the data in a fashion similar to the plane simulation.

The numerical results can be found in table B.2. We are listing the differences between the ground truth center point of the sphere and its estimated position first in each component x, y, z and then their distance from each other $\|\Delta x\|$. We also list the absolute error in the radius estimation Δr.

All methods cause an error in the order of millimeters in the location of the sphere and the radius. The radius is therefore measured up to roughly 2 % precision. The high uncertainty comes from the small measurement volume: Only a very small part of the sphere could be reconstructed due to the high convexity and small screen. It is very hard to properly fit a sphere into only a small fraction of it.

Figure 6.1 shows the reconstruction results for the NC method. White points are outside of the sphere, black points inside it. The hole effect discussed in section 4.3.2 is visible with white points floating above the sphere and black points dropping into it. It can also be seen that the different measurement pairs have a slightly different overlapping patch on the surface. The three-ways-normal comparison $NC(\mathcal{D}_0, \mathcal{D}_1, \mathcal{D}_2)$ reconstructs the surface in the complete volume, i.e. it combines the two patches to one. But otherwise it does not improve the reconstruction a lot: it inherits the strong hole problem of the first measurement and the numerical results are not significantly improved either.

Figure 6.2 shows the reconstruction results for the other methods. It is nice to see that that the Combi reconstructor significantly improves the PRGM result: while the latter shows the hole effect very strongly and drifts under and over the surface, the former has much less points inside the surface and the remaining ones lie very nicely on top of the surface. The numerical results are not improved much

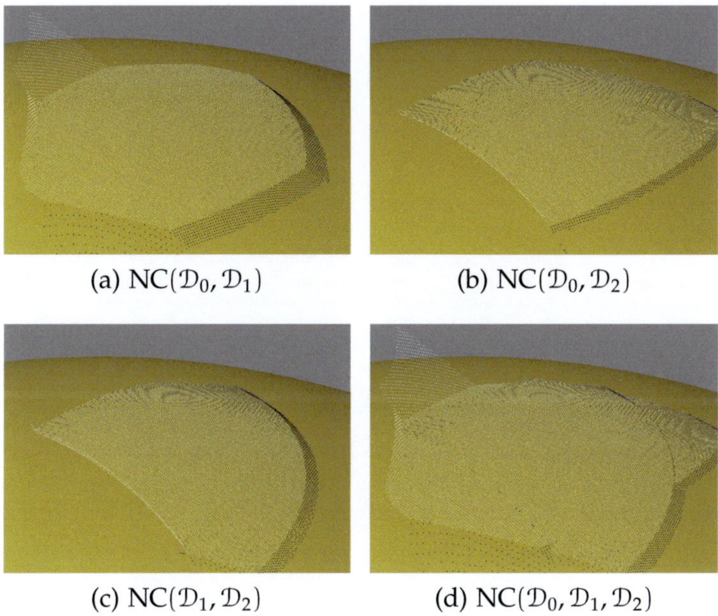

(a) $NC(\mathcal{D}_0, \mathcal{D}_1)$ (b) $NC(\mathcal{D}_0, \mathcal{D}_2)$

(c) $NC(\mathcal{D}_1, \mathcal{D}_2)$ (d) $NC(\mathcal{D}_0, \mathcal{D}_1, \mathcal{D}_2)$

Figure 6.1: Reconstruction results on the simulated sphere using the NC method. White points are outside the sphere, black points inside. One example for the hole effect can be seen in the top left image.

by combining NC and PRGM though, in some cases they even get slightly worse. The CR only shows minor drifting on the small reconstructed patch. It is also smooth except for some points on the right which suddenly drop inside the sphere. Numerically it is also on par with the other techniques for this experiment.

The Combi reconstructor performs best in this case. Its numerical results are among the best from all results and the reconstructed points lie well on the surface.

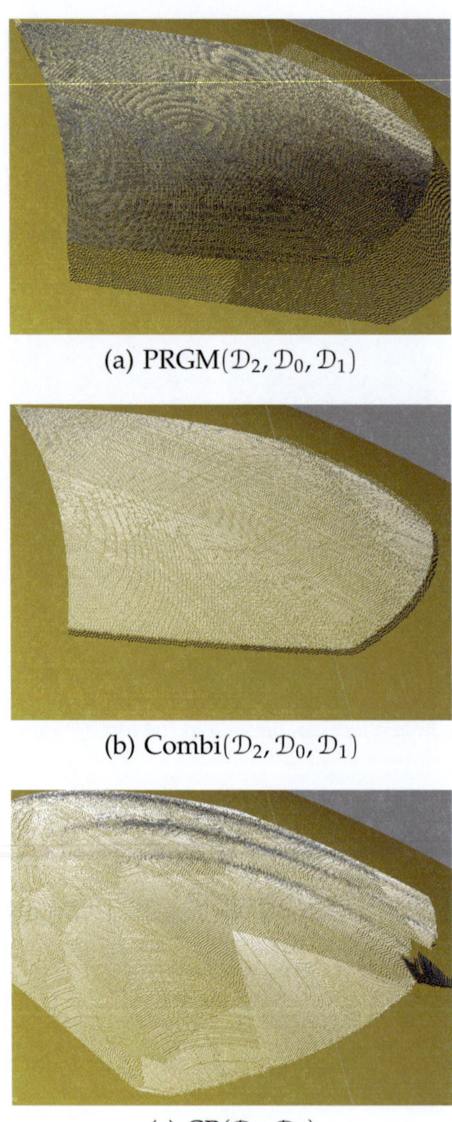

(a) $\text{PRGM}(\mathcal{D}_2, \mathcal{D}_0, \mathcal{D}_1)$

(b) $\text{Combi}(\mathcal{D}_2, \mathcal{D}_0, \mathcal{D}_1)$

(c) $\text{CR}(\mathcal{D}_1, \mathcal{D}_2)$

Figure 6.2: Representative results of the other reconstruction techniques on the simulated sphere.

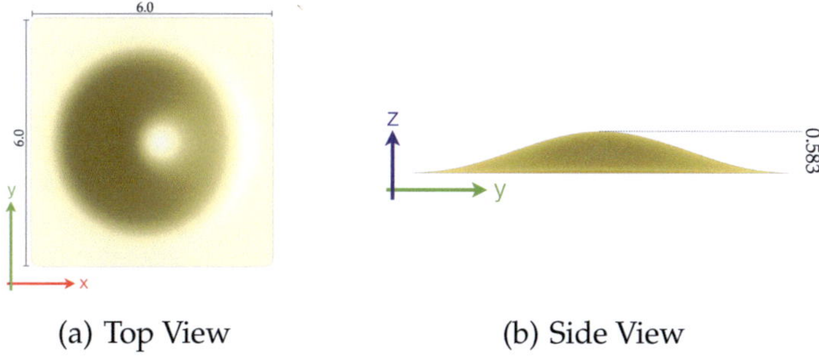

(a) Top View

(b) Side View

Figure 6.3: The simulated object "Hubbel"

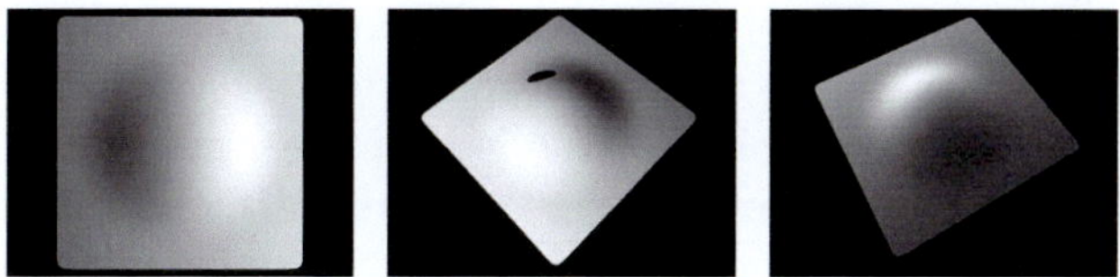

Figure 6.4: Masked results of the SGMF simulations of the Hubbel. The image in the middle shows the spot where the screen was not reflected – this effectively represents a hole in the measurement.

6.1.4 Hubbel

Our final reconstruction object is a free formed surface formed from a plane by embossing a „Hubbel" (German for bump) onto it. It is designed to be a more realistic model of real world objects. Its size is 6×6 meters and its height is roughly .6 meters in its center. It is symmetric around two axes and convex but its form does not follow a simple analytical expression. Its material is modelled to be highly polished gold. A top and side view can be seen in Figure 6.3.

As before, we took three measurements of the Hubbel, but this time, we also included one measurement with an artificial hole ($\mathcal{D}_1$), i.e. an area that did not reflect the screen because the screen was poorly positioned. In our case, this spot reflects the surroundings of the measurement. The three SGMF can be seen in Figure 6.4.

As the Hubbel cannot easily be captured analytically and because the raytracer subdivided the surface internally to make the rendering smoother, we cannot compare the quantitative results of the methods. We will only discuss the qualitative results and the impact the non-reflective hole has on the reconstruction.

The results for the NC method can be seen in Figure 6.5. The topmost result shows the reconstruction on two measurements without holes. As expected, the result looks flawless – all points are perfectly situated on top of the surface.

The image in the middle shows the reconstruction using the simulation with the hole and one without. At the edges of the hole, the effect predicted in section 4.3.2 can clearly be seen. Otherwise, the reconstruction is as good as the first one – the hole only has local effects on the quality of the reconstruction. The last result is using all three measurements and looks similar to the result of the first reconstruction. The hole has been filled because only the data of the first and third measurement where used inside. Only at the edges of the hole is the reconstruction slightly flawed: the border effect of the second measurement impacted some of these points so that they are not perfectly aligned on the surface. This is hardly noticeable in the pictures though.

Figure 6.6 shows the results for the PRGM and for the Combi reconstructor. For the PRGM, there are a lot more stray points that were not filtered out. Both reconstructors do not show the hole effect in the middle picture. This is in accordance with our other experiments: the PRGM and Combi reconstructor do not suffer from the hole effect as much as the NC and very often they do not show it at all for the same camera and screen positions at which the NC shows it. The PRGM pictures seems to contain vertical lines – this is just a random artifact though that appeared when points where removed for the visualization which is not in the three-dimensional data.

The PRGM closes the hole if all three measurements are used, but the reconstruction is flawed and far from the correct surface. Even though two of the measurements have valid data for the hole, the third measurement destroys a good reconstruction. The Combi reconstruction looks better overall: There are less stray points and in

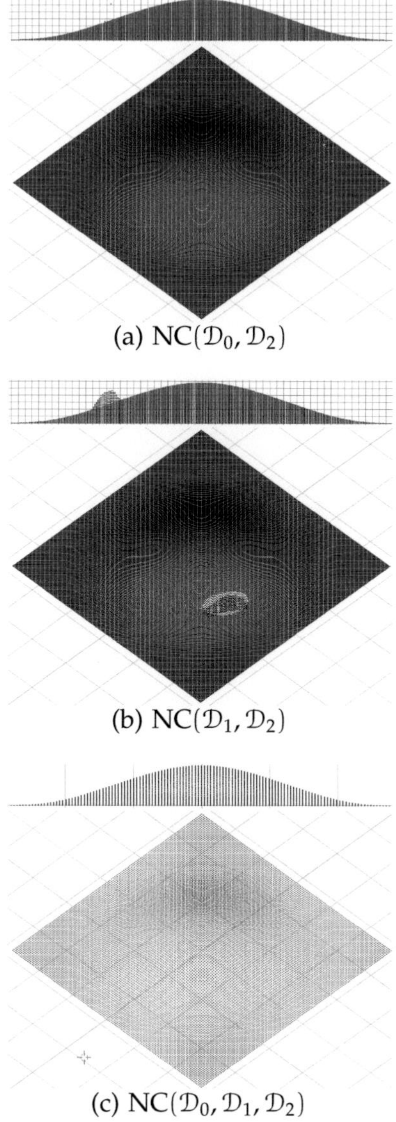

(a) $NC(\mathcal{D}_0, \mathcal{D}_2)$

(b) $NC(\mathcal{D}_1, \mathcal{D}_2)$

(c) $NC(\mathcal{D}_0, \mathcal{D}_1, \mathcal{D}_2)$

Figure 6.5: NC results on Hubbel.

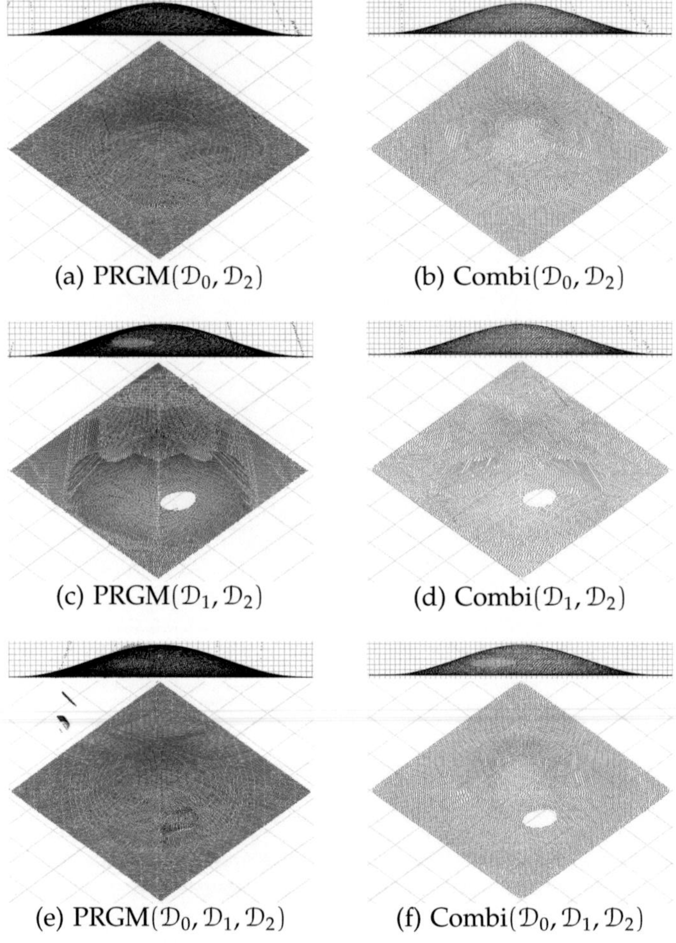

(a) PRGM$(\mathcal{D}_0, \mathcal{D}_2)$

(b) Combi$(\mathcal{D}_0, \mathcal{D}_2)$

(c) PRGM$(\mathcal{D}_1, \mathcal{D}_2)$

(d) Combi$(\mathcal{D}_1, \mathcal{D}_2)$

(e) PRGM$(\mathcal{D}_0, \mathcal{D}_1, \mathcal{D}_2)$

(f) Combi$(\mathcal{D}_0, \mathcal{D}_1, \mathcal{D}_2)$

Figure 6.6: PRGM and PRGM+NC results on Hubbel.

the three measurement reconstruction there are even none. The hole is not filled, but the hole effect is also completely filtered out – and no data is better than wrong data.

Figure 6.7 shows the results for the consistency reconstructor. The reconstruction with the measurements without holes (a) is looking nice, but a closeup reveals the drift error in the integration: the reconstructed points (white) are systematically below the surface. This is due to the integration drift that is inherent to this method. The other two pictures show the asymmetry of the method: the middle picture was reconstructed from a measurement without holes and the measurement with holes was for consistency testing. The measurement looks fine except for the drift. The other way around shows problems around the hole and at the very edge of the surface (c). Bad data in the first measurement therefore really makes more trouble than in the second.

Overall, given three measurements, the NC deals best with holes in the data. It closes the hole with the information from the other measurements. Otherwise, the Combi reconstructor is most convincing in this simulation: It has very little stray points and detects the hole problem and delivers no data in this case. The CR's integration drift hits here full force and disqualifies the CR practically for correct reconstruction. It is the only method that can smoothly close the hole with only two measurements though.

6.1.4.1 Testing the Auto-Calibration

To test the auto-calibration algorithms, we used the following test case: we disturbed the position and orientation of the screen in $\mathcal{D}_0$ and $\mathcal{D}_2$ by a random translation of $1\,\mathrm{cm}$ and rotation of $2.5\,^\circ$. We also disturbed the view direction of the camera in $\mathcal{D}_2$ by an angle of $2\,^\circ$. We then tested the auto-calibration algorithms described in chapter 5.

The NC auto-calibration sometimes drifted off and didn't terminate. When it converged, it found the correct minima. The PRGM and Combi reconstructor always converged on the correct minima, the Combi method consistently needed less iterations. The precision of

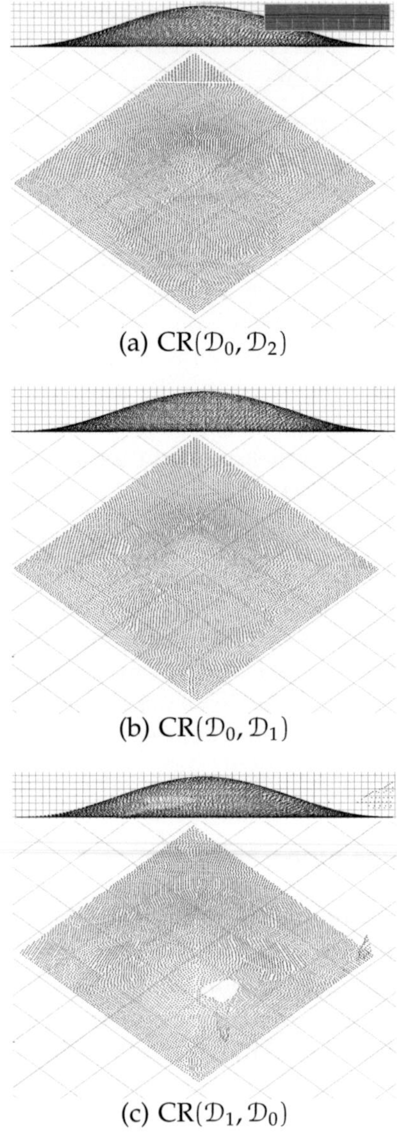

(a) $CR(\mathcal{D}_0, \mathcal{D}_2)$

(b) $CR(\mathcal{D}_0, \mathcal{D}_1)$

(c) $CR(\mathcal{D}_1, \mathcal{D}_0)$

Figure 6.7: CR results on Hubbel. The top image shows the drift in a zoomed area: the white points are the reconstruction results, the black lines are the true surface.

the final result of all three algorithms was only limited by the running time of the optimization. The runtime required for optimizing the extrinsics is in the order of a thousand reconstructions.

6.2 Experiments

We will now provide an evaluation of the reconstruction algorithms on real data. All data was acquired on a custom built experimental setup which we will discuss in detail in the next section. After that, we will provide information about the calibration of this experimental setup. Then we will discuss the experiments conducted with this setup and the results the algorithms produced.

6.2.1 Experimental Setup

The schematics and the basic working principle of the experimental setup can be seen in Figure 6.8, a photo of the real setup is provided in Figure 6.9. The setup consists of the screen - a standard LCD computer color display which is only driven with grayscale values, an industrial video camera capable of picturing images at 1600×1200 pixels mounted on a six degrees-of-freedom industrial robot arm and of course an object under test – the surface. The robot is capable of delivering reasonably precise position and orientation information of his hand – the manufacturers manual claims .2 mm precise position and .5 ° precise angle information. Together with a good hand-eye calibration – we will discuss this topic in section 6.2.1.2 – this directly translates into the extrinsics of the camera. However, our screen is hand positioned and there is no precise position data available. We know its dimensions though, and together with the camera data this is enough to get an estimate of the screen's position in the world coordinate frame (see section 6.2.1.4).

An experiment therefore needs two steps:

1. After the screen has been positioned manually, the camera is positioned such that it can picture the whole screen's surface.

(a) First step of measurement: Using the given position of the robot, the hand-eye calibration between robot and camera, the dimension of the screen and an image of the pattern on the screen, the screen's position and orientation is estimated.

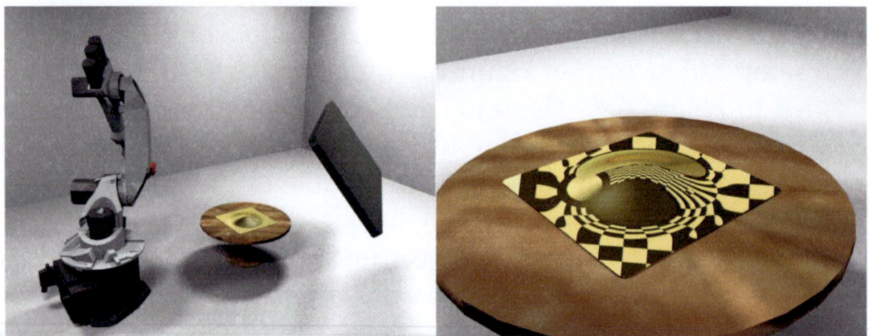

(b) Second step of measurement: Without moving the screen, the camera is moved to a position where all or most of the surface reflects the screen into the camera. A SGMF measurement is then run. Together with screen position and camera intrinsics and extrinsics this forms a deflectometric measurement.

Figure 6.8: Schematics and principle of the experimental setup. Images on the right are from the camera.

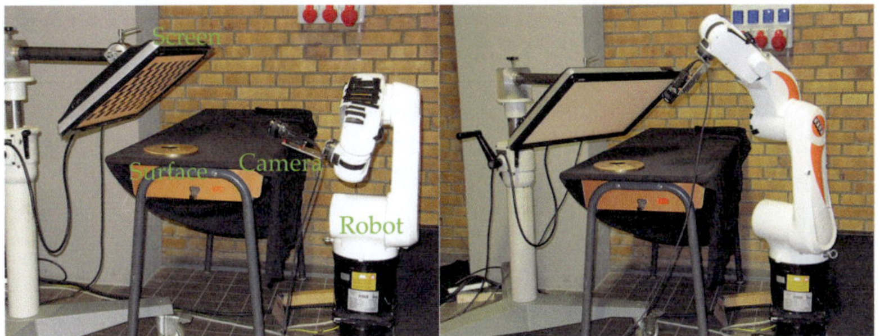

Figure 6.9: Photo of the experimental setup. The left picture shows the first step: the determination of the screen position. The right picture is taken while the deflectometric measurement is acquired.

> A chessboard pattern of known dimension is displayed and is used to determine the screen's position and orientation in the world coordinate frame.

2. The camera is then moved to a position where it can picture the reflected and distorted screen in the surface under test. A SGMF measurement can then be acquired. Together with the camera parameters and screen position, this represents a complete deflectometric measurement.

There are plenty of parameters to be calibrated in this setup before a measurement can be acquired. The most important one is the rigid transformation between robot and camera coordinate frame. This problem is well known as hand-eye calibration and its quality is of utmost importance in our measurement process. We will discuss the solutions we investigated in section 6.2.1.2. Other calibration parameters are the camera intrinsics (see section 6.2.1.1) and the screen's non linearity (see section 6.2.1.3).

6.2.1.1 Intrinsic Camera Calibration

We use a well established model and its corresponding calibration procedure [Zha00, HS97] for our camera. For all algorithms working with the acquired images, we assume a pinhole camera. For this to be valid, we need to compensate for the effects of the lens on the image data.

Given a point $\mathbf{x}$ which we will augment to its dual quaternion representation $\breve{\mathbf{x}} = (1,\mathbf{0}) + \epsilon\,(0,\mathbf{x})$, the pinhole camera model will project it into the pixel $\mathbf{u} = (u_x, u_y)^{\mathsf{T}}$ according to

$$
s \begin{pmatrix} u_x \\ u_y \\ 1 \end{pmatrix} = \mathbf{A}\mathrm{Vec}\left(\mathrm{Dual}\left(\breve{\mathbf{e}}\,\breve{\mathbf{x}}\,\overline{\breve{\mathbf{e}}}^{=}\right)\right).
\tag{6.2}
$$

Herein $\breve{\mathbf{e}}$ is the rigid body transformation between the calibration target and the camera coordinate frame. The matrix

$$
\mathbf{A} = \begin{pmatrix} f_x & \alpha f_x & c_x \\ 0 & f_y & c_y \\ 0 & 0 & 1 \end{pmatrix}
\tag{6.3}
$$

is a projective matrix containing the intrinsic parameters of the camera. These are the principal point (c_x, c_y) which is approximately the image center, the focal lengths f_x and f_y and the skew α which contains information about how strongly the x and y axes in the pixel plane deviate from $90\,°$. For todays industrial cameras, the skew is very close to zero and can be usually completely ignored.

A real life lens usually has some distortions. These are modeled using the radial coefficients k_1, k_2, k_3 and the tangential coefficients p_1, p_2. The complete transformation then becomes a little more complex:

$$
\begin{pmatrix} x \\ y \\ z \end{pmatrix} = \mathrm{Vec}\left(\mathrm{Dual}\left(\breve{\mathbf{e}}\,\breve{\mathbf{x}}\,\overline{\breve{\mathbf{e}}}^{=}\right)\right)
\tag{6.4}
$$

$$
x' = x/z \qquad y' = y/z
\tag{6.5}
$$

$$
r := x'^2 + y'^2
\tag{6.6}
$$

$$
u_x = f_x\left(x'(1 + k_1 r^2 + k_2 r^4 + k_3 r^6) + 2p_1 x' y' + p_2(r^2 + 2x'^2)\right) + c_x
\tag{6.7}
$$

$$
u_y = f_y\left(x'(1 + k_1 r^2 + k_2 r^4 + k_3 r^6) + 2p_2 x' y' + p_1(r^2 + 2y'^2)\right) + c_y.
\tag{6.8}
$$

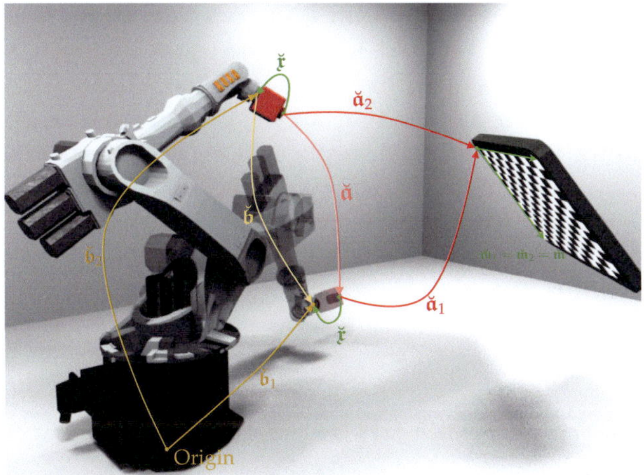

Figure 6.10: The various frames of references and their relationship in the hand-eye calibration process.

For the calibration we proceed as follows: we acquire an image of a chessboard of a certain size (e.g. with 6×8 squares, i.e. 35 internal corners). We change the position of the camera before each capture, i.e. we have different extrinsics for each acquired image of the chessboard. Our model has 5 parameters for the distortions, 5 parameters for the intrinsics and 6 parameters for each camera position, i.e. $10 + 6n$ parameters when n images with different camera positions are acquired. We can now use a bundle adjustment to solve for the parameters we are looking for. The metric that is usually employed for the minimization is the reprojection error of the chessboard corners through the model. We used the implementation suggested in [HS97] for all of our experiments.

6.2.1.2 Hand-Eye Calibration

The frame of reference of the camera and the frame of reference of the robot's hand coordinate differ and the transformation between those two systems must be calibrated in one way or another.

Using the identifiers from Figure 6.10, we search for the transformation $\breve{x}$ from hand to camera frame. The usual approach is to take some arm positions $\breve{b}_i$ relative to the world coordinate frame where the camera can see a calibration target (in our case a chessboard pattern). If the intrinsics and distortions of the camera are known, the screw that translates the frame of reference of the chessboard $\breve{m}_i$ into the frame of reference of the camera $\breve{a}_i$ can be determined. Between two such positions with $i = 1, 2$, the following relation can be read from Figure 6.10.

$$\breve{b} = \breve{x}\breve{a}\overline{\breve{x}} \tag{6.9}$$

where $\breve{a} = \overline{\breve{a}_1}\breve{a}_2$ is first going the transformation a_1 backwards, than doing a_2. Analogously, $\breve{b} = \breve{b}_1\overline{\breve{b}_2}$.

We will discuss three solutions for this problem in increasing sophistication and quality.

6.2.1.2.1 Linear Separable Solution Writing (6.9) using homogeneous matrices as introduced in section 1.4 the problem reads

$$\mathbf{X}^{-1}\mathbf{A}\mathbf{X} = \mathbf{B} \tag{6.10}$$
$$\Rightarrow \mathbf{A}\mathbf{X} = \mathbf{X}\mathbf{B}. \tag{6.11}$$

The various transformations are now represented by affine transformation matrices; the letters have been kept the same: $\mathbf{A}$ is the transformation between the two camera positions, $\mathbf{B}$ the one between robot hand positions and $\mathbf{X}$ is the transformation between hand and eye. We will also define that the transformation $\mathbf{A}$ consists of the rotation matrix $\mathbf{R}_A$ and the translational vector $\mathbf{t}_A$ and analogously for the other transformations.

This matrix equation can now be broken down in its rotational and its translational parts:

$$\mathbf{R}_A \mathbf{R}_X = \mathbf{R}_X \mathbf{R}_B \tag{6.12}$$

$$\mathbf{R}_A \mathbf{t}_X + \mathbf{t}_A = \mathbf{R}_X \mathbf{t}_B + \mathbf{t}_X$$
$$\Rightarrow (\mathbf{R}_A - \mathbf{I})\, \mathbf{t}_X = \mathbf{R}_X \mathbf{t}_B - \mathbf{t}_A \tag{6.13}$$

The rotational equation (6.12) can be further simplified using some of the properties of rotational matrices discussed in section 1.3.1, namely that each rotation matrix has an eigenvalue $\lambda = 1$ and that $\mathbf{R}^{-1} = \mathbf{R}^\mathsf{T}$. If we multiply the equation from the right site with the eigenvector $\mathbf{n}_B$ from $\mathbf{R}_B$ corresponding to its eigenvalue of 1, we see that

$$\mathbf{R}_A \mathbf{R}_X \mathbf{n}_B = \mathbf{R}_X \mathbf{R}_B \mathbf{n}_B = \mathbf{R}_X 1 \cdot \mathbf{n}_B. \tag{6.14}$$

From this, we can directly infer the eigenvector

$$\mathbf{n}_A = \mathbf{R}_X \mathbf{n}_B \tag{6.15}$$

of $\mathbf{R}_A$ to the eigenvalue 1. This equation is equivalent to (6.12), but much cheaper to compute.

We see from (6.13) that finding $\mathbf{t}_X$ becomes a linear problem if we have a solution for the rotational part $\mathbf{R}_X$. Of course, those two quantities are coupled. But for small rotations, first solving for $\mathbf{R}_X$ and then for $\mathbf{t}_X$ will yield acceptable results. The solution for the separable problem was first introduced by [FH86]; we follow a more compact mathematical formulation similar to [HD95].

So let's find a solution that minimizes the error in the rotational part (6.15). We start by writing the error function. For convenience, we write the rotation by $\mathbf{R}_x$ as a quaternion product with a unit quaternion $\mathfrak{r}$ and make some simplifications

$$\|\mathbf{n}_A - \mathfrak{r}\mathbf{n}_B\bar{\mathfrak{r}}\|^2 = \|\mathbf{n}_A - \mathfrak{r}\mathbf{n}_B\bar{\mathfrak{r}}\|^2\|\mathfrak{r}\|^2 \tag{6.16}$$

$$= \|\mathbf{n}_A\mathfrak{r} - \mathfrak{r}\mathbf{n}_B\|^2 \tag{6.17}$$

$$= (\mathbf{L}(\mathbf{n}_A)\mathbf{x} - \mathbf{R}(\mathbf{n}_B)\mathbf{x})^\mathsf{T}(\mathbf{L}(\mathbf{n}_A)\mathbf{x} - \mathbf{R}(\mathbf{n}_B)\mathbf{x}) \tag{6.18}$$

$$= \mathbf{x}^\mathsf{T}\mathbf{K}\mathbf{x}. \tag{6.19}$$

We used the identities (1.23) in the third step of this simplification – this also converts the quaternion to a 4-dimensional vector of real values – and defined

$$\mathbf{K} := (\mathbf{L}(\mathbf{n}_A) - \mathbf{R}(\mathbf{n}_B))^\mathsf{T}(\mathbf{L}(\mathbf{n}_A) - \mathbf{R}(\mathbf{n}_B)) \tag{6.20}$$

in the last step. We must now minimize (6.16) under the constraint that $\|\mathbf{x}\| = 1$. We can achieve this reduction in the degrees of freedom by using a Lagrange multiplier and instead minimize

$$\min_{\mathbf{x}} \left(\mathbf{x}^{\mathsf{T}} \mathbf{K} \mathbf{x} + \lambda(1 - \mathbf{x}^{\mathsf{T}} \mathbf{x}) \right) \tag{6.21}$$

This convex equation always has a unique global minimum which can be directly found by differentiating and setting equal to zero. This gives

$$\mathbf{K} \mathbf{x} = \lambda \mathbf{x}. \tag{6.22}$$

The quaternion that minimizes the rotational error (6.12) is therefore the eigenvector of $\mathbf{K}$ to its smallest eigenvalue. As $\mathbf{K}$ is positive and symmetric, the eigenvalues will all be real and positive. Solving for the translation $\mathbf{t}_X$ is then a simple linear least squares problem.

Note that data from many movements can and should be combined into a big composed matrix $\mathbf{K}_{\text{total}} = \sum \mathbf{K}_i$ which is equivalent to minimizing the sum of squared differences over all movement rotation errors under the constraint of a unit quaternion. Of course, the same applies to the solution for the translational part.

6.2.1.2.2 Coupled solution For larger rotations, the decoupled solution delivers unsatisfying results. An easy approach is to simply minimize a coupled error function like the following

$$\min_{\underline{r},\mathbf{t}_x} \left(\lambda_1 \|\mathbf{n}_a - \underline{r} \mathbf{n}_B \bar{\underline{r}}\|^2 + \lambda_2 \|(\mathbf{R}_A - \mathbf{I})\mathbf{t}_X - \underline{r}\mathbf{t}_B\bar{\underline{r}} + \mathbf{t}_A\|^2 \right). \tag{6.23}$$

This minimizer function was suggested by [HD95], but they used a Levenberg-Marquardt minimizer and a soft constraint on the unit length of the rotational quaternion. In our experiments, we used a superior solver that can handle constraints directly called SLSQP (Sequential least squares fitting with constraints) which is well described in [CC09].

6.2.1.2.3 Linear solution using Dual Quaternions In [Dan99] a nice method of solving the hand-eye calibration procedure is described which directly uses the dual quaternion formulation. We follow the description by beginning with the original transformation equation (6.9) and note that the scalar part of the equation does not contribute to the solution because it is the same for $\breve{a}$ and $\breve{b}$:

$$\mathrm{Sc}\,(\breve{b}) = \frac{1}{2}(\breve{b} + \overline{\breve{b}}) = \frac{1}{2}(\breve{r}\breve{a}\overline{\breve{r}} + \overline{\breve{r}\breve{a}\overline{\breve{r}}})$$
$$= \frac{1}{2}\breve{r}(\breve{a} + \overline{\breve{a}})\overline{\breve{r}} = \mathrm{Sc}\,(\breve{a})\,\breve{r}\overline{\breve{r}} = \mathrm{Sc}\,(\breve{a})\,. \tag{6.24}$$

This is the screw congruence theorem which states that two rigidly connected bodys which undergo a rigid body motion together will make both a screw motion with the same angle and the same pitch but different rotation axes relative to their own frame of reference. Or in other words: the pitch and the angle of a screw remain invariant under rigid transformations [Che91].

Since the scalar components do not give us any more information they cancel from the equation and we can set them to zero right away and ignore them in the further analysis.

We now split (6.9) into the dual and non-dual part and remember that $\breve{a} = (0, \mathbf{a}_r) + \epsilon\,(0, \mathbf{a}_d)$ and likewise for $\breve{b}$.

$$\mathbf{b}_r = \mathbf{r}_r\mathbf{a}_r\overline{\mathbf{r}}_r \tag{6.25}$$
$$\mathbf{b}_d = \mathbf{r}_r\mathbf{a}_r\overline{\mathbf{r}}_d + \mathbf{r}_r\mathbf{a}_d\overline{\mathbf{r}}_r + \mathbf{r}_d\mathbf{a}_r\overline{\mathbf{r}}_r \tag{6.26}$$

Multiplication with $\mathbf{r}_r$ from the right and using the dual normalization relationship for unit dual quaternions

$$\overline{\mathbf{r}}_r\mathbf{r}_d + \overline{\mathbf{r}}_d\mathbf{r}_r = 0 \tag{6.27}$$

and also inserting the first equation into the second gives

$$\mathbf{b}_r\mathbf{r}_r = \mathbf{r}_r\mathbf{a}_r \tag{6.28}$$
$$\mathbf{b}_d\mathbf{r}_r = -\mathbf{b}_r\mathbf{r}_d + \mathbf{r}_r\mathbf{a}_d + \mathbf{r}_d\mathbf{a}_r \tag{6.29}$$

or equivalently

$$\mathbf{b}_r\mathbf{r}_r - \mathbf{r}_r\mathbf{a}_r = 0 \tag{6.30}$$
$$(\mathbf{b}_d\mathbf{r}_r - \mathbf{r}_r\mathbf{a}_d) + (\mathbf{b}_r\mathbf{r}_d - \mathbf{r}_d\mathbf{a}_r) = 0. \tag{6.31}$$

We now translate this equation in a vector-matrix product following the rules of the quaternion multiplication and keeping in mind that the scalar parts of all quaternions are 0. This gives the following equation:

$$\begin{pmatrix} \mathbf{a_r} - \mathbf{b_r} & [\mathbf{a_r} + \mathbf{b_r}]_\times & 0 & 0 \\ \mathbf{a_d} - \mathbf{b_d} & [\mathbf{a_d} + \mathbf{b_d}]_\times & \mathbf{a_r} - \mathbf{b_r} & [\mathbf{a_r} + \mathbf{b_r}]_\times \end{pmatrix} \begin{pmatrix} \mathbf{x_r} \\ \mathbf{x_d} \end{pmatrix} = \mathbf{0}. \qquad (6.32)$$

Note that the matrix corresponds to one hand-eye movement, if we have many movements, we construct a matrix with all information by stacking the individual 6×8 matrices. This matrix has a rank of six in the noise-free case. We therefore find a basis for the nullspace of the matrix B which has dimension 2 by singular value decomposition. We then use the normalization constraints for the unit quaternions to find the two valid dual quaternions $\check{\mathbf{r}}$ and $\bar{\check{\mathbf{r}}}$, any of which is a solution to our hand-eye calibration problem.

6.2.1.2.4 Discussion In our experiments, the linear separable solution proved unusable. The rotational part between camera and robot is quite large and the separation approximation was just not valid.

The coupled solution and the dual quaternion solution gave plausible and similar results – with the running times of the dual quaternion method being much faster than the coupled approach. Which result is more precise cannot be determined without ground truth data, but we have a higher confidence in the dual quaternion data – mainly based on the results of [Che91], but also because there are only six free parameters in this solution instead of seven. We therefore used the last method for our hand-eye calibration.

6.2.1.3 Photometric Calibration

We already discussed in section 3.2.4.2 what influence the non linearities of the screen's response can have on the SGMF. Obviously, this directly translates into errors in the reconstruction. To solve this problem, we tried two approaches. We already described the first one in section 3.2.4.2. The response curve of our screen can be approximated very well by a second order polynomial (see Figure 3.2) given the viewing angles are not too big. That is, the analytical solution for the phase shift should not be biased in our case, though the solutions for the amplitude and the background illuminations will be.

To also measure them without bias and to not limit ourselves to too small viewing angles, we tried a second solution using a per-pixel lookup table for each measurement. Before we started a SGMF measurement, we displayed ten pictures of increasing brightness on the screen. We used the corresponding camera pictures to fit a per-pixel spline lookup curve into these measurements and used the inverse mapping as a lookup table which linearizes the response curve. This solution worked quite acceptably for the brighter part of the lookup table but gave noisy data in the darker part. Since the final results for the SMGF were noisier than with the analytical solution, we discarded the lookup table approach and only used the analytical approach in our measurements.

6.2.1.4 Determining the Screen Position

If the transformation $\check{r}$ between hand and eye is known, the screen position is easily found. We display a chessboard of known size on the screen and picture it with the camera mounted on the robot. By minimizing the reprojection error, we can find the transformation $\check{a}$ from the camera to the screen. We now only need the transformation $\check{b}$ from world to robot arm frame and the hand-eye transformation $\check{r}$ and we directly get the transformation $\check{w}_s$ from the screen coordinate frame to the world, i.e. the extrinsics of the screen as

$$\check{w}_s = \overline{\check{b}}\,\check{r}\,\overline{\check{a}}. \tag{6.33}$$

Geometrically, a flat angle would be ideal for this approach. However, extracting the corners of the checkerboard works best when the camera looks along the normal vector of the screen and will become harder when the camera looks in a flat angle onto the surface. A compromise must be struck here.

The edges of the chessboard will be found more precisely if the chessboard is bigger – the distance to the screen will therefore also have an impact on the precision. But even in ideal conditions, the position of the screen will not be very precise with the error being in the order of a few millimeters. This is due to the detection of the chessboard corners which is very sensitive to tilting and rotating. The precision could be improved by taking more than one image with different camera positions into account.

This concludes the description of the experimental setup. We will now discuss the experiments conducted with it.

6.2.2 Mirror

The first test object is a round, flat mirror with a radius of 5 cm which can be seen in Figure 6.11(a). As in section 6.1.2, we can do a numerical analysis of the methods here – we simply fit a plane using RANSAC to the measured points and compare the mean distance of the points from the plane which were not classified as outliers. Due to the imperfection of our calibration data, we must always start the test of our algorithms with the auto-calibration. This is a necessity for all conducted experiments and we will not be able to separate the steps. Therefore, we will only consider the algorithms that can provide auto-calibration together with reconstruction, namely PRGM and Combi reconstructors and we will always discuss the combined result of auto-calibration and reconstruction method in this chapter.

We ran the following experiments: We used the PRGM and the Combi auto-calibration procedure and varied only the screens of the measurements (S1 S2), the screens and the second camera (S1 S2 C2) or all four elements (S1 S2 C1 C2). The numerical results can be seen in table B.3. The S1 S2 C2 auto-calibrations perform consistently worse than S1 S2 and S1 S2 C1 C2 which seems strange at first. It seems that the optimization is not able to drag the complete coordinate system into the frame of reference of the first camera. The reconstruction considering all extrinsics as free parameters performs best considering all cases – but they also take a long time to compute. On our setup, a reconstruction plus auto-calibration with 24

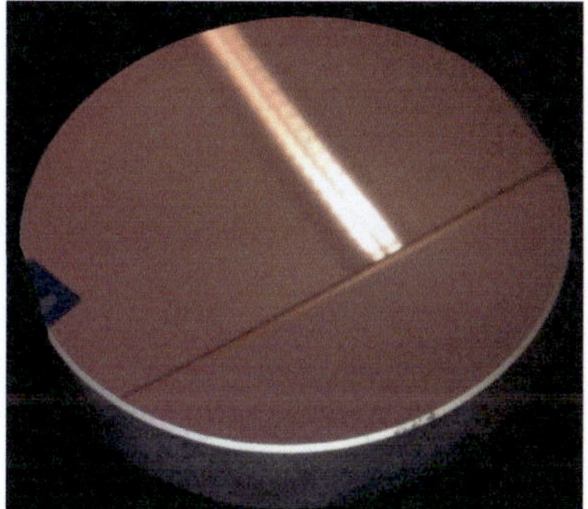

(a) The flat mirror reflecting an office ceiling.

(b) The bowling ball.

Figure 6.11: Objects used in the experiments.

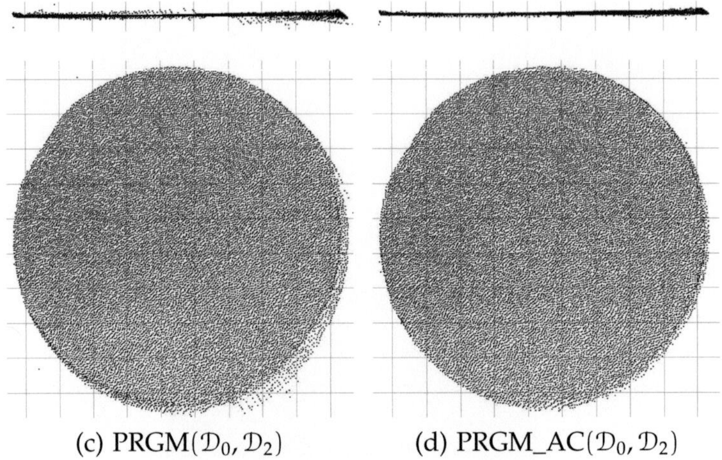

(c) $\text{PRGM}(\mathcal{D}_0, \mathcal{D}_2)$ (d) $\text{PRGM\_AC}(\mathcal{D}_0, \mathcal{D}_2)$

Figure 6.12: Reconstructed points from the mirror before (left) and after (right) auto-calibration.

free parameters ran for approximately three hours. It is notable that the PRGM outperforms the Combi reconstructor here in all cases.

Figure 6.12 compares the results of a PRGM auto-calibration with a measurement without auto-calibration. It is obvious that the auto-calibrated measurement has less outliers, though some remain at the edges of the surface.

6.2.3 Bowling Ball

In our second experiment, we used a bowling ball with a radius of $r = 109.15\,\text{mm}$. We took two deflectometric measurements of the object, auto-calibrated and reconstructed, filtered the results and used RANSAC to fit a model of a sphere onto the points. The results can be seen in table B.4.

Both methods improve the results comparably, again the calibration that takes all objects as free parameters performs the best. The Combi reconstructor outperforms the PRGM in this test slightly. The final relative error for the radius is below 1%. The experiment therefore has better results than the simulation. This is because the ratio of screen size and sphere radius is smaller here. Therefore, a single measurement can picture a larger part of the sphere which makes it easier to fit the data to the points.

6.2.4 Faucet

Our last and most challenging object is a brass faucet with some holes in it. The exact size of the object is unknown. It is meant as a challenging object to push the methods to their limits. It is approximately 20 cm long, 17 cm wide, and 1 cm thick. It also contains a number of dents and scratches.

The object itself and the auto-calibration results of PRGM and Combo can be seen in Figure 6.13. A result of the Combo reconstructor without auto-calibration can be seen in Figure 6.14. Clearly, the auto-calibration significantly improves the result. However, the number of outliers remains high even after auto-calibration. The circles in Figure 6.13 indicate dents in the surface: they are barely visible in the photo, but can be seen as areas without proper reconstruction in the result images. The Combo reconstructor is able to close some of the areas better than the PRGM and it also performs slightly better in capturing the geometry of the object.

Both methods have a lot of problems with the holes: the hole effect is visible in the side view images and there are a lot of outliers in and around the holes. The final calibration is obviously not perfect because we can see artifacts around the bumps that are similar to what we could see in the simulated image in section 5.1.

6.3 Conclusion

The simulations and experiments show that the best methods for reconstruction are the PRGM and the Combi reconstructor. The CR has the problem of global drift which makes it a bad contender overall. The NC performs well in situations where the extrinsics are well

Figure 6.13: Reconstruction with auto-calibration using PRGM (left) and Combo (right). The object seems truncated at the bottom left, because one of the measurements did not deliver data there.

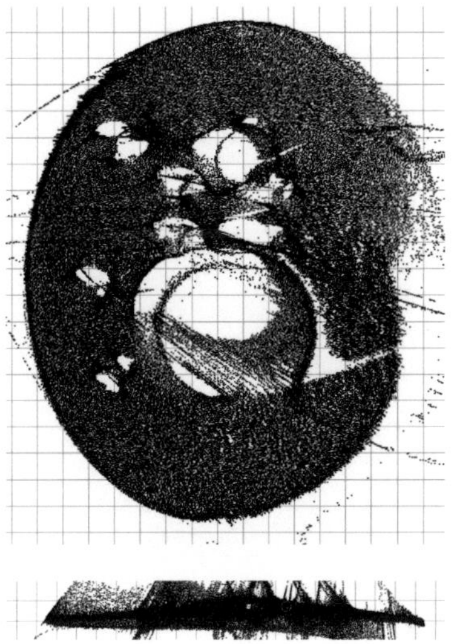

Figure 6.14: $\text{Combo}(\mathcal{D}_0, \mathcal{D}_1)$

known but proves unstable when used with the auto-calibration. PRGM and Combi are performing very well and are comparable if good extrinsics are available but the Combi reconstructor is slightly more stable when used in situations without proper extrinsic calibration. However, it is numerically more expensive than the PRGM.

Chapter 7

Summary and Outlook

This work describes mathematics, techniques and experimental setup suitable for reconstructing highly specular reflecting surfaces using only auto-calibrating deflectometry.

We started out by introducing the mathematical concepts needed for the reconstruction. We proceeded by explaining the basic principle and some theoretical limits of deflectometry. We also introduced the fundamental formulas using the mathematical framework of dual quaternions.

We detailed the common acquisition methods for the SGMF and explained our methods of choice. We used the Multi-Phase Shift method for our experiments. After the acquisition of the SGMF, we were in a position to discuss the reconstruction algorithms. We presented established and novel algorithms and detailed their mode of operation. We also stated them explicitly in the context of the mathematical framework. Due to the high performance of the GPU implementation, we were in a position to invent various combinations of the algorithms with each other and generalizations to more than two measurements. Both extensions have shown to improve the results in most cases. All algorithms have been tested on simulations and on experimental data.

The experimental setup with the camera mounted on a robot and the screen being free and without positioning information inspired the search for a auto-calibrating approach: while the robot delivers data with an acceptable precision, the localisation of the screen is not very precise. This was also seen in the experimental evaluation. However, the auto-calibration approach using the PRGM or the Combi reconstructor improved the reconstruction results considerably. Altogether, the Combi reconstructor has proven to be the best all around algorithm for reconstruction and auto-calibration. It dealt well with holes in the measurement and outliers and produced the best results in the real world experiments and the simulations.

Future work could include improvements of the experimental setup. The robot has a high repeatability, but the absolute positioning is not known precisely. If absolute positioning could be guaranteed, the setup could be used as a benchmark for the auto-calibration. The experimental data could also be improved by replacing the video camera currently deployed by a high quality still camera which might improve the acquired image quality considerably.

Another open question this work unraveled is the relationship between camera and screen position and the hole problem. Some camera-screen position pairs reduce the hole problem at the border of a surface considerably compared to others. This specific observation can be generalized to the question if there are ideal or degenerate position pairs or sets of deflectometric measurements. It is also of interest if given an object, it is possible to find a set of screen and camera pairs that capture the whole surface in an ideal fashion.

This directly leads to another open question. Given a convex or even a cylindrical object for reconstruction, a flat screen is a handicap: it must be very big to guarantee that most or all of the surface reflects light from it into the camera. This can be solved in two ways: either by allowing for a curved screen or by patching more measurements together to form a bigger virtual surface. We will discuss both possibilities now.

A curved screen would solve many problems with convex objects. Organic LED technology is able to provide a curved screen with suitable characteristics for deflectometry. The position detection from section 6.2.1.4 would need to be changed and the measurement of the SGMF would also need to be updated for a new screen geometry. The reconstruction algorithms would still work as stated.

The other possibility is to study the possibilities of patching many measurements of the current setup into a combined surface. Of course, combining three-dimensional point clouds is a well-researched topic, but for deflectometry an earlier fusion of the SGMF is already possible. It has yet to be studied which approach will lead to better results.

The auto-calibration provided in this work is already useful in improving reconstruction results. It was made possible in part due to the significant improvements of the speed of reconstruction we achieved in this work. However, the reconstruction with auto-calibration is still rather slow. Further work on the algorithms could try improving the speed and the quality of the final extrinsics.

The bottom line of this thesis is that a simple and dynamic experimental setup as used in this work together with a set of algorithms for reconstruction and calibration provides a flexible toolchain for the measurement of specular surfaces. The auto-calibration is slow but functional and has made manual calibration unnecessary while improving the final results.

Appendix **A**

Geometrical Relations

A.1 Distance between two Lines

Given two lines in normalized Plücker coordinates $\check{l}_1 = \mathbf{u}_1 + \epsilon\, \mathbf{v}_1$ and $\check{l}_2 = \mathbf{u}_2 + \epsilon\, \mathbf{v}_2$ we find the smallest signed distance d between them by constructing a unit vector that is perpendicular to both

$$\mathbf{n} = \mathbf{u}_1 \times \mathbf{u}_2. \tag{A.1}$$

and projecting the difference vector of any two points on the line onto $\mathbf{n}$. Points on a line are parametrisized via $\mathbf{u} \times \mathbf{v} + t\mathbf{u}$ with $t \in \mathbb{R}$. We can choose $t = 0$ which gives.

$$d = \mathbf{n}(\mathbf{u}_1 \times \mathbf{v}_1 - \mathbf{u}_2 \times \mathbf{v}_2) \tag{A.2}$$

$$= (\mathbf{u}_1 \times \mathbf{u}_2)(\mathbf{u}_1 \times \mathbf{v}_1) - (\mathbf{u}_1 \times \mathbf{u}_2)(\mathbf{u}_2 \times \mathbf{v}_2) \tag{A.3}$$

$$= \begin{vmatrix} \mathbf{u}_1 \cdot \mathbf{u}_1 & \mathbf{u}_1 \cdot \mathbf{v}_1 \\ \mathbf{u}_2 \cdot \mathbf{u}_1 & \mathbf{u}_2 \cdot \mathbf{v}_1 \end{vmatrix} - \begin{vmatrix} \mathbf{u}_1 \cdot \mathbf{u}_2 & \mathbf{u}_1 \cdot \mathbf{v}_2 \\ \mathbf{u}_2 \cdot \mathbf{u}_2 & \mathbf{u}_2 \cdot \mathbf{v}_2 \end{vmatrix} \tag{A.4}$$

$$= \begin{vmatrix} 1 & 0 \\ \mathbf{u}_2 \cdot \mathbf{u}_1 & \mathbf{u}_2 \cdot \mathbf{v}_1 \end{vmatrix} - \begin{vmatrix} \mathbf{u}_1 \cdot \mathbf{u}_2 & \mathbf{u}_1 \cdot \mathbf{v}_2 \\ 1 & 0 \end{vmatrix} \tag{A.5}$$

$$= \mathbf{u}_1\mathbf{v}_2 + \mathbf{u}_2\mathbf{v}_1 \tag{A.6}$$

Here we used the Lagrange Identity

$$(\mathbf{a} \times \mathbf{b})(\mathbf{c} \times \mathbf{d}) = \begin{vmatrix} \mathbf{a} \cdot \mathbf{c} & \mathbf{a} \cdot \mathbf{d} \\ \mathbf{b} \cdot \mathbf{c} & \mathbf{b} \cdot \mathbf{d} \end{vmatrix} \tag{A.7}$$

We now look at the dot product between the two lines

$$\check{\mathbf{l}}_1 \cdot \check{\mathbf{l}}_2 = \mathbf{u}_1 \cdot \mathbf{u}_2 + \epsilon \left(\mathbf{u}_1 \mathbf{v}_2 + \mathbf{u}_2 \mathbf{v}_1 \right) \tag{A.8}$$

$$= \|\mathbf{u}_1\|\|\mathbf{u}_2\| \cos \Theta + \epsilon\, d \tag{A.9}$$

$$= \cos \Theta + \epsilon\, d, \tag{A.10}$$

and note that the distance between the two lines d appears naturally as the dual part of the result.

A.2 Plane-Line meet

Given is the plane $\mathbf{p} = (d, \mathbf{n})$ and a line not inside this plane $\check{\mathbf{l}} = \mathbf{l} + \epsilon\, \mathbf{m}$. We are interested in their point of intersection $\mathbf{p}$. The equation of the plane is

$$\mathbf{n} \cdot \mathbf{x} + d = 0. \tag{A.11}$$

Using a parametric equation for the line with the parameter t

$$\mathbf{x} = \mathbf{m} \times \mathbf{l} + t\mathbf{l}, \tag{A.12}$$

inserting this into the plane equation and solving for t yields

$$t = \frac{-\mathbf{n} \cdot (\mathbf{m} \times \mathbf{l}) - d}{\mathbf{n} \cdot \mathbf{l}}. \tag{A.13}$$

We can now resubstitute into the line equation:

$$\mathbf{x} = \mathbf{m} \times \mathbf{l} + \frac{-\mathbf{n} \cdot (\mathbf{m} \times \mathbf{l}) - d}{\mathbf{n} \cdot \mathbf{l}}\mathbf{l} \tag{A.14}$$

$$= \frac{(\mathbf{n} \cdot \mathbf{l})(\mathbf{m} \times \mathbf{l}) - (\mathbf{n} \cdot (\mathbf{m} \times \mathbf{l}))\mathbf{l} - d\mathbf{l}}{\mathbf{n} \cdot \mathbf{l}} \tag{A.15}$$

$$= \frac{\mathbf{n} \times (-\mathbf{l} \times (\mathbf{m} \times \mathbf{l})) - d\mathbf{l}}{\mathbf{n} \cdot \mathbf{l}} \tag{A.16}$$

$$= \frac{\mathbf{m} \times \mathbf{n} - d\mathbf{l}}{\mathbf{n} \cdot \mathbf{l}} \tag{A.17}$$

We needed the relation (1.66) twice, first from left to right. Then to find

$$-\mathbf{l} \times (\mathbf{m} \times \mathbf{l}) = (-\mathbf{l} \cdot \mathbf{l})\mathbf{m} - (-\mathbf{l} \cdot \mathbf{m})\mathbf{l} = -\mathbf{m}. \tag{A.18}$$

A.3 Reflection as Dual Quaternion Operation

We will use the definitions from section 2.3 in this calculation. First, we will look at the real part of (2.12):

$$\mathbf{n}_l \mathbf{s}_l \mathbf{n}_l = \left(-\mathbf{n} \cdot \mathbf{s}, \mathbf{n} \times \mathbf{s}\right)\mathbf{n}_l \tag{A.19}$$

$$= (0, (\mathbf{n} \times \mathbf{s}) \times \mathbf{n} - (\mathbf{ns})\mathbf{n}) \tag{A.20}$$

$$= (0, -(\mathbf{n} \cdot \mathbf{s})\mathbf{n} + (\mathbf{n} \cdot \mathbf{n})\mathbf{s} - (\mathbf{n} \cdot \mathbf{s})\mathbf{n}) \tag{A.21}$$

$$= (0, \mathbf{s} - 2(\mathbf{n} \cdot \mathbf{s})\mathbf{n}) \tag{A.22}$$

We used again the relation (1.66) and that $\mathbf{n}$ has length 1.

The first term in the dual part is very similar and becomes

$$\mathbf{n}_l \mathbf{s}_m \mathbf{n}_l = \left(0, \mathbf{s}_m - 2(\mathbf{n} \cdot \mathbf{s}_m)\mathbf{n}\right). \tag{A.23}$$

The second term in the dual part is

$$\mathbf{n}_l \mathbf{s}_l \mathbf{n}_l = (-\mathbf{n} \cdot \mathbf{s}, \mathbf{n} \times \mathbf{s})\mathbf{n}_m \tag{A.24}$$

$$= \left(-(\mathbf{n} \times \mathbf{s})(\mathbf{n} \times \mathbf{p}), -(\mathbf{n} \cdot \mathbf{s})(\mathbf{n} \times \mathbf{p}) + (\mathbf{n} \times \mathbf{s}) \times (\mathbf{n} \times \mathbf{p})\right) \tag{A.25}$$

$$= \left(-\begin{vmatrix} \mathbf{n} \cdot \mathbf{n} & \mathbf{n} \cdot \mathbf{p} \\ \mathbf{s} \cdot \mathbf{n} & \mathbf{s} \cdot \mathbf{p} \end{vmatrix}, -(\mathbf{n} \cdot \mathbf{s})(\mathbf{n} \times \mathbf{p}) + (\mathbf{n} \cdot (\mathbf{s} \times \mathbf{p}))\mathbf{n}\right) \tag{A.26}$$

$$= \left(-\mathbf{s} \cdot \mathbf{p} + (\mathbf{n} \cdot \mathbf{p})(\mathbf{n} \cdot \mathbf{s}), -(\mathbf{n} \cdot \mathbf{s})(\mathbf{n} \times \mathbf{p}) + (\mathbf{n} \cdot (\mathbf{s} \times \mathbf{p}))\mathbf{n}\right) \tag{A.27}$$

$$= \left(-\mathbf{s} \cdot \mathbf{p} + (\mathbf{n} \cdot \mathbf{p})(\mathbf{n} \cdot \mathbf{s}), -(\mathbf{n} \cdot \mathbf{s})(\mathbf{n} \times \mathbf{p}) + (\mathbf{s} \cdot (\mathbf{p} \times \mathbf{n}))\mathbf{n}\right). \tag{A.28}$$

We used the Lagrange identity (A.7) again. We also used

$$(\mathbf{a} \times \mathbf{b}) \times (\mathbf{a} \times \mathbf{c}) = (\mathbf{a} \cdot (\mathbf{b} \times \mathbf{c}))\mathbf{a}. \qquad (A.29)$$

and that the triple vector product is invariant to cyclic permutations

$$(\mathbf{a} \times \mathbf{b}) \cdot \mathbf{c} = \mathbf{a} \cdot (\mathbf{b} \times \mathbf{c}). \qquad (A.30)$$

The third and final term in the dual part is

$$\mathbf{n}_m \mathbf{s}_l \mathbf{n}_l = \Big(-(\mathbf{n} \times \mathbf{p}) \cdot \mathbf{s}, (\mathbf{n} \times \mathbf{p}) \times \mathbf{s} \Big)\mathbf{n}_l \qquad (A.31)$$

$$= \Big(\mathbf{n} \cdot (\mathbf{s} \times (\mathbf{n} \times \mathbf{p})), \mathbf{s} \cdot (\mathbf{p} \times \mathbf{n})\mathbf{n} - ((\mathbf{s} \times (\mathbf{n} \times \mathbf{p})) \times \mathbf{n}) \Big) \qquad (A.32)$$

$$= \Big((\mathbf{s} \cdot \mathbf{p}) - (\mathbf{s} \cdot \mathbf{n})(\mathbf{n} \cdot \mathbf{p}), (\mathbf{s} \cdot (\mathbf{p} \times \mathbf{n}))\mathbf{n} + (\mathbf{s} \cdot \mathbf{n})(\mathbf{p} \times \mathbf{n}) \Big). \qquad (A.33)$$

Summing the terms up reveals a scalar part of zero. The vector part becomes

$$\mathbf{r}_m = \mathbf{s}_m + 2\big((\mathbf{n} \cdot \mathbf{s})(\mathbf{p} \times \mathbf{n}) + (\mathbf{s} \cdot (\mathbf{p} \times \mathbf{n}) - \mathbf{s}_m \cdot \mathbf{n})\mathbf{n}\big). \qquad (A.34)$$

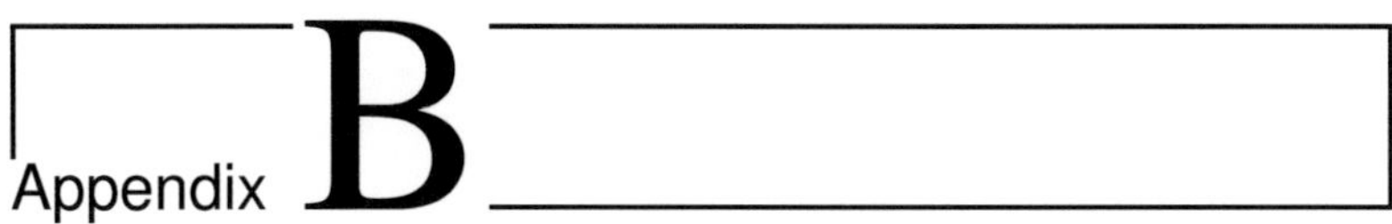

Appendix B

Results of Experiments

This appendix lists the result of all methods on the data presented in section 6.1 and section 6.2. The discussion of the data is done in the corresponding sections in the main text.

Table B.1: Results of reconstruction on simulated plane patch (see section 6.1.2).

Method	Arguments	n_x	n_y	n_z	d	Δ_{mean}
NC	$\mathcal{D}_0\ \mathcal{D}_1$	-7.32×10^{-6}	-2.98×10^{-6}	1.00	-9.91×10^{-5}	9.90×10^{-5}
NC	$\mathcal{D}_0\ \mathcal{D}_2$	1.31×10^{-5}	-9.88×10^{-7}	1.00	-9.62×10^{-5}	9.70×10^{-5}
NC	$\mathcal{D}_1\ \mathcal{D}_2$	2.67×10^{-7}	-2.10×10^{-7}	1.00	-6.77×10^{-5}	6.78×10^{-5}
NC	$\mathcal{D}_0\ \mathcal{D}_1\ \mathcal{D}_2$	1.06×10^{-6}	-1.49×10^{-6}	1.00	-8.03×10^{-5}	8.03×10^{-5}
PRGM	$\mathcal{D}_0\ \mathcal{D}_1$	-5.87×10^{-6}	-3.62×10^{-6}	1.00	-1.01×10^{-4}	1.01×10^{-4}
PRGM	$\mathcal{D}_0\ \mathcal{D}_2$	1.13×10^{-5}	-6.42×10^{-6}	1.00	-1.02×10^{-4}	1.02×10^{-4}
PRGM	$\mathcal{D}_1\ \mathcal{D}_2$	2.69×10^{-6}	-2.91×10^{-6}	1.00	-7.49×10^{-5}	7.48×10^{-5}
PRGM	$\mathcal{D}_0\ \mathcal{D}_1\ \mathcal{D}_2$	4.63×10^{-6}	-5.63×10^{-6}	1.00	-9.18×10^{-5}	9.23×10^{-5}
PRGM	$\mathcal{D}_1\ \mathcal{D}_2\ \mathcal{D}_0$	-1.20×10^{-6}	-4.79×10^{-6}	1.00	-8.46×10^{-5}	8.29×10^{-5}
PRGM	$\mathcal{D}_2\ \mathcal{D}_0\ \mathcal{D}_1$	9.17×10^{-8}	1.78×10^{-6}	1.00	-6.71×10^{-5}	6.76×10^{-5}
Combi	$\mathcal{D}_0\ \mathcal{D}_1$	-7.01×10^{-6}	-3.33×10^{-6}	1.00	-1.00×10^{-4}	9.95×10^{-5}
Combi	$\mathcal{D}_0\ \mathcal{D}_2$	1.36×10^{-5}	-4.86×10^{-6}	1.00	-1.04×10^{-4}	1.06×10^{-4}
Combi	$\mathcal{D}_1\ \mathcal{D}_2$	2.48×10^{-6}	-2.30×10^{-6}	1.00	-7.45×10^{-5}	7.36×10^{-5}
Combi	$\mathcal{D}_0\ \mathcal{D}_1\ \mathcal{D}_2$	-1.58×10^{-6}	4.12×10^{-6}	-1.00	9.16×10^{-5}	9.16×10^{-5}
Combi	$\mathcal{D}_1\ \mathcal{D}_2\ \mathcal{D}_0$	-7.29×10^{-7}	-3.54×10^{-6}	1.00	-8.30×10^{-5}	8.15×10^{-5}
Combi	$\mathcal{D}_2\ \mathcal{D}_0\ \mathcal{D}_1$	9.79×10^{-7}	1.81×10^{-6}	1.00	-7.09×10^{-5}	7.19×10^{-5}
CR	$\mathcal{D}_0\ \mathcal{D}_1$	1.02×10^{-7}	3.10×10^{-6}	1.00	-1.71×10^{-4}	1.71×10^{-4}
CR	$\mathcal{D}_1\ \mathcal{D}_0$	1.75×10^{-4}	6.24×10^{-4}	1.00	-3.82×10^{-3}	5.88×10^{-3}
CR	$\mathcal{D}_0\ \mathcal{D}_2$	1.06×10^{-7}	3.10×10^{-6}	1.00	-9.07×10^{-5}	9.07×10^{-5}
CR	$\mathcal{D}_2\ \mathcal{D}_0$	-9.67×10^{-4}	-6.32×10^{-4}	1.00	-1.28×10^{-3}	2.95×10^{-3}

Method	Arguments	n_x	n_y	n_z	d	Δ_{mean}
CR	$\mathcal{D}_1\,\mathcal{D}_2$	6.88×10^{-4}	-9.02×10^{-4}	-1.00	-6.95×10^{-4}	9.66×10^{-4}
CR	$\mathcal{D}_2\,\mathcal{D}_1$	-9.84×10^{-6}	6.27×10^{-6}	1.00	-2.62×10^{-4}	2.67×10^{-4}

Table B.2: Results of reconstruction on a simulated sphere (see section 6.1.3).

Method	Arguments	Δx	Δy	Δz	$\lVert \Delta \mathbf{x} \rVert$	Δ_r
NC	$\mathcal{D}_0\,\mathcal{D}_1$	-1.66×10^{-4}	6.60×10^{-4}	1.61×10^{-3}	1.75×10^{-3}	1.85×10^{-3}
NC	$\mathcal{D}_0\,\mathcal{D}_2$	1.81×10^{-4}	2.39×10^{-5}	-3.87×10^{-3}	3.87×10^{-3}	3.75×10^{-3}
NC	$\mathcal{D}_1\,\mathcal{D}_2$	3.79×10^{-4}	-2.99×10^{-4}	-6.99×10^{-3}	7.00×10^{-3}	6.85×10^{-3}
NC	$\mathcal{D}_0\,\mathcal{D}_1\,\mathcal{D}_2$	3.96×10^{-4}	-1.07×10^{-4}	-3.62×10^{-3}	3.64×10^{-3}	3.49×10^{-3}
PRGM	$\mathcal{D}_0\,\mathcal{D}_1$	-1.17×10^{-3}	-3.99×10^{-5}	-8.41×10^{-3}	8.49×10^{-3}	7.89×10^{-3}
PRGM	$\mathcal{D}_0\,\mathcal{D}_2$	-8.16×10^{-4}	2.62×10^{-4}	-6.13×10^{-3}	6.19×10^{-3}	5.84×10^{-3}
PRGM	$\mathcal{D}_1\,\mathcal{D}_2$	-3.62×10^{-4}	-6.69×10^{-4}	-7.05×10^{-3}	7.09×10^{-3}	6.88×10^{-3}
PRGM	$\mathcal{D}_0\,\mathcal{D}_1\,\mathcal{D}_2$	-9.54×10^{-4}	-1.96×10^{-4}	-9.75×10^{-3}	9.79×10^{-3}	9.41×10^{-3}
PRGM	$\mathcal{D}_1\,\mathcal{D}_2\,\mathcal{D}_0$	-5.52×10^{-4}	-2.72×10^{-4}	-2.73×10^{-3}	2.80×10^{-3}	2.60×10^{-3}
PRGM	$\mathcal{D}_2\,\mathcal{D}_0\,\mathcal{D}_1$	6.92×10^{-4}	-7.68×10^{-4}	2.33×10^{-4}	2.55×10^{-3}	2.11×10^{-3}
Combi	$\mathcal{D}_0\,\mathcal{D}_1$	-5.69×10^{-6}	4.06×10^{-4}	-3.49×10^{-4}	5.35×10^{-4}	1.18×10^{-4}
Combi	$\mathcal{D}_0\,\mathcal{D}_2$	5.81×10^{-5}	9.36×10^{-5}	-3.30×10^{-3}	3.31×10^{-3}	3.18×10^{-3}
Combi	$\mathcal{D}_1\,\mathcal{D}_2$	2.38×10^{-4}	-2.51×10^{-4}	-6.51×10^{-3}	6.52×10^{-3}	6.36×10^{-3}
Combi	$\mathcal{D}_0\,\mathcal{D}_1\,\mathcal{D}_2$	8.92×10^{-5}	1.35×10^{-4}	-2.35×10^{-3}	2.35×10^{-3}	2.21×10^{-3}
Combi	$\mathcal{D}_1\,\mathcal{D}_2\,\mathcal{D}_0$	1.69×10^{-3}	-2.05×10^{-3}	-9.39×10^{-3}	9.76×10^{-3}	9.27×10^{-3}

Method	Arguments	Δx	Δy	Δz	$\|\Delta \mathbf{x}\|$	Δ_r
Combi	$\mathcal{D}_2\,\mathcal{D}_0\,\mathcal{D}_1$	5.40×10^{-5}	8.68×10^{-5}	-3.52×10^{-3}	3.53×10^{-3}	3.41×10^{-3}
CR	$\mathcal{D}_0\,\mathcal{D}_1$	1.46×10^{-3}	-3.44×10^{-4}	4.47×10^{-4}	1.56×10^{-3}	5.27×10^{-4}
CR	$\mathcal{D}_0\,\mathcal{D}_2$	8.97×10^{-4}	1.16×10^{-3}	3.85×10^{-3}	4.12×10^{-3}	3.97×10^{-3}
CR	$\mathcal{D}_1\,\mathcal{D}_2$	7.25×10^{-4}	1.00×10^{-3}	-1.74×10^{-3}	2.13×10^{-3}	1.52×10^{-3}

Table B.3: Auto-calibration and reconstruction on a flat mirror (see section 6.2.2).

Method	Arguments	n_x	n_y	n_z	d	Δ_{mean}
PRGM	$\mathcal{D}_0\,\mathcal{D}_1$ S1 S2	-1.80×10^{-2}	1.64×10^{-2}	1.00	5.74×10^2	2.45×10^{-1}
PRGM	$\mathcal{D}_0\,\mathcal{D}_1$ S1 S2 C2	-2.65×10^{-2}	7.88×10^{-3}	1.00	5.76×10^2	3.65×10^{-1}
PRGM	$\mathcal{D}_0\,\mathcal{D}_1$ S1 S2 C1 C2	-2.88×10^{-3}	4.86×10^{-3}	1.00	5.82×10^2	1.42×10^{-1}
PRGM	$\mathcal{D}_0\,\mathcal{D}_2$ S1 S2	-2.61×10^{-2}	-1.20×10^{-2}	1.00	5.83×10^2	1.85×10^{-1}
PRGM	$\mathcal{D}_0\,\mathcal{D}_2$ S1 S2 C2	-2.57×10^{-2}	2.11×10^{-3}	1.00	5.77×10^2	2.62×10^{-1}
PRGM	$\mathcal{D}_0\,\mathcal{D}_2$ S1 S2 C1 C2	-1.29×10^{-2}	-8.52×10^{-3}	1.00	5.86×10^2	1.75×10^{-1}
Combi	$\mathcal{D}_0\,\mathcal{D}_1$ S1 S2	-1.71×10^{-2}	1.58×10^{-2}	1.00	5.74×10^2	2.34×10^{-1}
Combi	$\mathcal{D}_0\,\mathcal{D}_1$ S1 S2 C2	-2.60×10^{-2}	7.86×10^{-3}	1.00	5.76×10^2	3.39×10^{-1}
Combi	$\mathcal{D}_0\,\mathcal{D}_1$ S1 S2 C1 C2	-5.40×10^{-3}	9.24×10^{-4}	1.00	5.83×10^2	1.45×10^{-1}
Combi	$\mathcal{D}_0\,\mathcal{D}_2$ S1 S2	-2.77×10^{-2}	-3.13×10^{-3}	1.00	5.79×10^2	1.90×10^{-1}
Combi	$\mathcal{D}_0\,\mathcal{D}_2$ S1 S2 C2	-1.77×10^{-2}	2.61×10^{-2}	1.00	5.67×10^2	2.26×10^{-1}
Combi	$\mathcal{D}_0\,\mathcal{D}_2$ S1 S2 C1 C2	-1.55×10^{-2}	-1.05×10^{-2}	1.00	5.87×10^2	1.92×10^{-1}

Table B.4: Auto-calibration on a bowling ball (see section 6.2.3).

Method	Arguments	c_x	c_y	c_z	Δ_r
PRMG	$\mathcal{D}_0\ \mathcal{D}_1$ S1 S2	5.61×10^2	-5.73×10^{-1}	6.88×10^{-1}	4.10×10^{-3}
PRMG	$\mathcal{D}_0\ \mathcal{D}_1$ S1 S2 C2	5.61×10^2	-5.72×10^{-1}	6.83×10^{-1}	9.38×10^{-4}
PRMG	$\mathcal{D}_0\ \mathcal{D}_1$ S1 S2 C1 C2	5.61×10^2	-5.72×10^{-1}	6.83×10^{-1}	8.70×10^{-4}
Combi	$\mathcal{D}_0\ \mathcal{D}_1$ S1 S2	5.61×10^2	-5.72×10^{-1}	6.88×10^{-1}	4.26×10^{-3}
Combi	$\mathcal{D}_0\ \mathcal{D}_1$ S1 S2 C2	5.62×10^2	-5.72×10^{-1}	6.82×10^{-1}	1.13×10^{-3}
Combi	$\mathcal{D}_0\ \mathcal{D}_1$ S1 S2 C1 C2	5.61×10^2	-5.72×10^{-1}	6.83×10^{-1}	8.30×10^{-4}

Bibliography

[Bal08]   Jonathan Balzer. *Regularisierung des Deflektometrieproblems.* PhD thesis, Universität Karlsruhe, Feb 2008.

[BHG⁺74]   JH Bruning, D.R. Herriott, JE Gallagher, DP Rosenfeld, AD White, and DJ Brangaccio. Digital wavefront measuring interferometer for testing optical surfaces and lenses. *Applied Optics*, 13(11):2693–2703, 1974.

[BR79]   O. Bottema and B. Roth. *Theoretical kinematics.* Dover Pubns, 1979.

[BWB06]   Jonathan Balzer, Stefan Werling, and Jürgen Beyerer. Regularization of the deflectometry problem using shading data. In Peisen S. Huang, editor, *Proceedings of the SPIE Optics East*, volume 6382, page 63820B. SPIE, 2006.

[CC09]   S.C. Chapra and R.P. Canale. *Numerical methods for engineers, 6th edition.* McGraw-Hill Science, 2009.

[Che91]   H.H. Chen. A screw motion approach to uniqueness analysis of head-eye geometry. In *Computer Vision and Pattern Recognition, 1991. Proceedings CVPR'91., IEEE Computer Society Conference on*, pages 145–151. IEEE, 1991.

[Cli71]   Clifford. Preliminary sketch of biquaternions. *Proceedings of the London Mathematical Society*, s1-4(1):381–395, 1871.

[Dan99]   K. Daniilidis. Hand-eye calibration using dual quaternions. *The International Journal of Robotics Research*, 18(3):286, 1999.

[Dem]   W. Demtröder. Experimentalphysik, volume 2: Elektrizität und optik.

[DF07]   L. Dorst and D. Fontijne. *Geometric algebra for computer science: an object-oriented approach to geometry*. Morgan Kaufmann, 2007.

[Dic89]   O.E. Dictionary. Oxford: Oxford university press, 1989.

[EKH07]   S. Ettl, J. Kaminski, and G. Häusler. Generalized hermite interpolation with radial basis functions considering only gradient data. In *Curve and Surface Fitting: Avignon 2006*, pages 141–149. Nashboro Press, 2007.

[Far99]   G.E. Farin. *NURBS: from projective geometry to practical use*. AK Peters, Ltd., 1999.

[FB81]   Martin A. Fischler and Robert C. Bolles. Random sample consensus: a paradigm for model fitting with applications to image analysis and automated cartography. *Commun. ACM*, 24(6):381–395, 1981.

[FC88]   R.T. Frankot and R. Chellappa. A method for enforcing integrability in shape from shading algorithms. *Pattern Analysis and Machine Intelligence, IEEE Transactions on*, 10(4):439–451, 1988.

[FH86]   O.D. Faugeras and M. Hebert. The representation, recognition, and locating of 3d shapes from range data. *Int'l J. Robotics Res*, 5(3):27–52, 1986.

[FPT10]   M. Fischer, M. Petz, and R. Tutsch. Bewertung von lcd-monitoren für deflektometrische messsysteme. *Sensoren und Messsysteme 2010*, 2010.

[FPT11]   M. Fischer, M. Petz, and R. Tutsch. Phasenrauschen in optischen messsystemen mit struktruierter beleuchtung. In F. Puente León and J. Beyerer, editors, *Messtechnisches Symposium der AHMT*, pages 115 – 125, 9 2011.

[Gal01]   J.H. Gallier. *Geometric Methods and Applications: For Computer Science and Engineering*. Texts in Applied Mathematics. Springer, 2001.

[GRA53]   F. GRAY. Pulse code communication, March 17 1953. US Patent 2,632,058.

[Häu99]   G. Häusler. Verfahren und vorrichtung zur ermittlung der form oder der abbildungseigenschaften von spiegelnden oder transparenten objekten. *Patent DE*, 19944354:A1, 1999.

[HD95]   R. Horaud and F. Dornaika. Hand-eye calibration. *The international journal of robotics research*, 14(3):195, 1995.

[HK05]   Jan W. Horbach and Soeren Kammel. Deflectometric inspection of diffuse surfaces in the far-infrared spectrum. In Jeffery R. Price and Fabrice Meriaudeau, editors, *Proceedings of SPIE*, volume 5679, pages 108–117. SPIE, 2005.

[Hor06]   C. Horneber. *Phasenmessende Deflektometrie - Ein Verfahren zur hochgenauen Vermessung spiegelnder Oberflächen*. PhD thesis, Universität Erlangen-Nürnberg, 2006.

[Hor07]   Jan Horbach. *Verfahren zur optischen 3D-Vermessung spiegelnder Oberflächen*. PhD thesis, Universität Karlsruhe, September 2007.

[HS93]   J.M. Huntley and H. Saldner. Temporal phase-unwrapping algorithm for automated interferogram analysis. *Applied Optics*, 32(17):3047–3052, 1993.

[HS97]   J. Heikkila and O. Silven. A four-step camera calibration procedure with implicit image correction. In *IEEE Computer Society Conference on Computer Vision and Pattern Recognition*, pages 1106–1112. INSTITUTE OF ELECTRICAL ENGINEERS INC (IEEE), 1997.

[HWB10]   Sebastian Höfer, Stefan Werling, and Jürgen Beyerer. Neuartige strategie zur vollständigen kalibrierung eines sensorsystems zur automatischen sichtprüfung spiegelnder oberflächen. In Fernando Puente León and Michael Heizmann, editors, *Forum Bildverarbeitung*, pages 25–34, Regensburg, December 2010. KIT Scientific Publishing.

[IH79]   K. Ikeuchi and B.K.P. Horn. *An application of the photometric stereo method*. 1979.

[Jäh95]   B. Jähne. *Digital image processing: concepts, algorithms and scientific applications*. Springer-Verlag London, UK, 1995.

[Jäh10]   B. Jähne. Emva 1288 standard for machine vision–objective specification of vital camera dat. *Optik & Photonik*, 1:53–54, 2010.

[JKW94]   B. Jähne, J. Klinke, and S. Waas. Imaging of short ocean wind waves: a critical theoretical review. *JOSA A*, 11(8):2197–2209, 1994.

[JSR05]   B. Jähne, M. Schmidt, and R. Rocholz. Combined optical slope/height measurements of short wind waves: principle and calibration. *Measurement Science and Technology*, 16:1937, 2005.

[Kam05]   Sören Kammel. *Deflektometrische Untersuchung spiegelnd reflektierender Freiformflächen*. Dissertation, Universität Karlsruhe (TH), Karlsruhe, 2005. Schriftenreihe Institut für Mess- und Regelungstechnik, Universitätsverlag Karlsruhe, Nr. 004.

[KH05]   Sören Kammel and Jan Horbach. Topography reconstruction of specular surfaces. In J.-Angelo Beraldin, Sabry F. El-Hakim, Armin Gruen, and James S. Walton, editors, *Videometrics VIII*, volume 5665, pages 59–66. SPIE, 2005.

[KLSS88]   K. Kinnstaetter, A.W. Lohmann, J. Schwider, and N. Streibl. Accuracy of phase shifting interferometry. *Applied Optics*, 27(24):5082–5089, 1988.

[Kna06]   Markus C. Knauer. *Absolute Phasenmessende Deflektometrie*. PhD thesis, Universität Erlangen-Nürnberg, Mai 2006.

[Knu68]   D. E. Knuth. *The Art of Computer Programming. Volume 1: Fundamental Algorithms*. Addison-Wesley, 1968.

[KPL+12]   Andreas· Klöckner, Nicolas· Pinto, Yunsup· Lee, B.· Catanzaro, Paul· Ivanov, and Ahmed Fasih. PyCUDA and PyOpenCL: A Scripting-Based Approach to GPU Run-Time Code Generation. *Parallel Computing*, 38(3):157–174, 2012.

[Kui99]   J.B. Kuipers. *Quaternions and rotation sequences*. Princeton university press Princeton, NJ, 1999.

[Lam03]   R. Lampalzer. *Physikalische und informationstheoretische Eigenschaften und Grenzen von Systemen zur optischen Formerfassung und Inspektion nach dem Prinzip der phasenmessenden Triangulation*. PhD thesis, Universität Erlangen-Nürnberg, 2003.

[M+09]   A. Munshi et al. The opencl specification. *Khronos OpenCL Working Group*, pages 11–15, 2009.

[NWT10]   Heung-Sun Ng, Tai-Pang Wu, and Chi-Keung Tang. Surface-from-gradients without discrete integrability enforcement: A gaussian kernel approach. *Pattern Analysis and Machine Intelligence, IEEE Transactions on*, 32(11):2085–2099, nov. 2010.

[Pér01]   Denis Pérard. *Automated Visual Inspection of Specular Surfaces with Structured-Lighting Reflection Techniques*. Dissertation, Universität Karlsruhe (TH), Karlsruhe, 2001. Fortschritt-Berichte, Reihe 8, Nr. 869, VDI-Verlag, Düsseldorf.

[PFT+86]   W.H. Press, B.P. Flannery, S.A. Teukolsky, W.T. Vetterling, et al. *Numerical recipes*, volume 547. Cambridge Univ Press, 1986.

[PR01]   Marcus Petz and Reinhold Ritter. Reflection grating method for 3d measurement of reflecting surfaces. In *Proceedings of SPIE*, volume 4399, pages 35–41. SPIE, 2001.

[PT04]   Marcus Petz and Rainer Tutsch. Rasterreflexionsphotogrammetrie zur messung spiegelnder oberflächen. *TM - Technisches Messen*, 71(7/8/2004):389–397, 2004.

[Rap07]   H. Rapp. Experimental and theoretical investigation of correlating TOF-camera systems. *University of Heidelberg*, 2007.

[Rat95]   C. Rathjen. Statistical properties of phase-shift algorithms. *JOSA A*, 12(9):1997–2008, 1995.

[RFHJ08]   H. Rapp, M. Frank, FA Hamprecht, and B. Jahne. A theoretical and experimental investigation of the systematic errors and statistical uncertainties of time-of-flight-cameras. *International Journal of Intelligent Systems Technologies and Applications*, 5(3):402–413, 2008.

[RS10]   H. Rapp and C. Stiller. Deflektometrische Methoden zur Sichtprüfung und 3D-Vermessung voll reflektierender Freiformflächen. In *Forum Bildverarbeitung 2010, Regensburg*, page 217. KIT Scientific Publishing, 2010.

[RS11]   H. Rapp and C. Stiller. Probabilistische Verbesserung der Kalibrierung deflektometrischer Systeme. In F. Puente León and J. Beyerer, editors, *Messtechnisches Symposium der AHMT*, pages 83 – 90, 9 2011.

[SCR99]   G. Sansoni, M. Carocci, and R. Rodella. Three-dimensional vision based on a combination of gray-code and phase-shift light projection: analysis and compensation of the systematic errors. *Applied Optics*, 38(31):6565–6573, 1999.

[SPB04]   J. Salvi, J. Pages, and J. Batlle. Pattern codification strategies in structured light systems. *Pattern Recognition*, 37(4):827–849, 2004.

[SS03]   R.S. Schone and O. Schwarz. Hybrid phase unwrapping algorithm extended by a minimum-cost-matching strategy. In *Proceedings of SPIE*, volume 4933, page 305, 2003.

[Stu91]   E. Study. Von den bewegungen und umlegungen. *Mathematische Annalen*, 39:441–566, 1891.

[Sur96]   Y. Surrel. Design of algorithms for phase measurements by the use of phase stepping. *Applied optics*, 35(1):51–60, 1996.

[Sur97]   Y. Surrel. Additive noise effect in digital phase detection. *Applied optics*, 36(1):271–276, 1997.

[TLGS05]   Marco Tarini, Hendrik P. A. Lensch, Michael Goesele, and Hans-Peter Seidel. 3d acquisition of mirroring objects using striped patterns. *Graphical Models*, 67(4):233–259, 2005.

[WB11]   S. Werling and J. Beyerer. Oberflächenrekonstruktion mittels Stereodeflektometrie. *tm-Technisches Messen*, 78(9):398–405, 2011.

[Wer11]   S.B. Werling. *Deflektometrie zur automatischen Sichtprüfung und Rekonstruktion spiegelnder Oberflächen*. KIT Scientific Publishing, 2011.

[WH03]   C. Wagner and G. Häusler. Information theoretical optimization for optical range sensors. *Applied optics*, 42(27):5418–5426, 2003.

[Yan02]   S.Y. Yan. *Number theory for computing*. Springer Verlag, 2002.

[Zha00]   Z. Zhang. A flexible new technique for camera calibration. *Pattern Analysis and Machine Intelligence, IEEE Transactions on*, 22(11):1330–1334, 2000.

[Zha09]   Song Zhang. Phase unwrapping error reduction framework for a multiple-wavelength phase-shifting algorithm. *Optical Engineering*, 48(10):105601, 2009.

[ZLY07]   Song Zhang, Xiaolin Li, and Shing-Tung Yau. Multi-level quality-guided phase unwrapping algorithm for real-time three-dimensional shape reconstruction. *Appl. Opt.*, 46(1):50–57, 2007.

[ZS95]   Bing Zhao and Yves Surrel. Phase shifting: six-sample self-calibrating algorithm insensitive to the second harmonic in the fringe signal. *Optical Engineering*, 34(9):2821–2822, 1995.

[ZSXS11]   Q. Zhang, X. Su, L. Xiang, and X. Sun. 3-d shape measurement based on complementary gray-code light. *Optics and Lasers in Engineering*, 2011.

[ZZB09]   K. Zhou, M. Zaitsev, and S. Bao. Reliable two-dimensional phase unwrapping method using region growing and local linear estimation. *Magnetic Resonance in Medicine*, 62(4):1085–1090, 2009.

**Schriftenreihe
Institut für Mess- und Regelungstechnik
Karlsruher Institut für Technologie
(1613-4214)**

Die Bände sind unter www.ksp.kit.edu als PDF frei verfügbar oder als
Druckausgabe bestellbar.

Band 001   Hans, Annegret
**Entwicklung eines Inline-Viskosimeters auf Basis eines
magnetisch-induktiven Durchflussmessers.** 2004
ISBN 3-937300-02-3

Band 002   Heizmann, Michael
**Auswertung von forensischen Riefenspuren mittels
automatischer Sichtprüfung.** 2004
ISBN 3-937300-05-8

Band 003   Herbst, Jürgen
**Zerstörungsfreie Prüfung von Abwasserkanälen mit
Klopfschall.** 2004
ISBN 3-937300-23-6

Band 004   Kammel, Sören
**Deflektometrische Untersuchung spiegelnd
reflektierender Freiformflächen.** 2005
ISBN 3-937300-28-7

Band 005   Geistler, Alexander
**Bordautonome Ortung von Schienenfahrzeugen mit
Wirbelstrom-Sensoren.** 2007
ISBN 978-3-86644-123-1

Band 006   Horn, Jan
**Zweidimensionale Geschwindigkeitsmessung
texturierter Oberflächen mit flächenhaften
bildgebenden Sensoren.** 2007
ISBN 978-3-86644-076-0

Band 007   Hoffmann, Christian
            **Fahrzeugdetektion durch Fusion monoskopischer
            Videomerkmale.** 2007
            ISBN 978-3-86644-139-2

Band 008   Dang, Thao
            Kontinuierliche Selbstkalibrierung von Stereokameras.
            2007
            ISBN 978-3-86644-164-4

Band 009   Kapp, Andreas
            **Ein Beitrag zur Verbesserung und Erweiterung der
            Lidar-Signalverarbeitung für Fahrzeuge.** 2007
            ISBN 978-3-86644-174-3

Band 010   Horbach, Jan
            **Verfahren zur optischen 3D-Vermessung spiegelnder
            Oberflächen.** 2008
            ISBN 978-3-86644-202-3

Band 011   Böhringer, Frank
            **Gleisselektive Ortung von Schienenfahrzeugen mit
            bordautonomer Sensorik.** 2008
            ISBN 978-3-86644-196-5

Band 012   Xin, Binjian
            **Auswertung und Charakterisierung dreidimensionaler
            Messdaten technischer Oberflächen mit Riefentexturen.**
            2009
            ISBN 978-3-86644-326-6

Band 013   Cech, Markus
            **Fahrspurschätzung aus monokularen Bildfolgen für
            innerstädtische Fahrerassistenzanwendungen.** 2009
            ISBN 978-3-86644-351-8

Band 014   Speck, Christoph
            **Automatisierte Auswertung forensischer Spuren auf
            Patronenhülsen.** 2009
            ISBN 978-3-86644-365-5

Band 015   Bachmann, Alexander
            Dichte Objektsegmentierung in Stereobildfolgen. 2010
            ISBN 978-3-86644-541-3

Band 016   Duchow, Christian
Videobasierte Wahrnehmung markierter Kreuzungen mit lokalem Markierungstest und Bayes'scher Modellierung. 2011
ISBN 978-3-86644-630-4

Band 017   Pink, Oliver
Bildbasierte Selbstlokalisierung von Straßenfahrzeugen. 2011
ISBN 978-3-86644-708-0

Band 018   Hensel, Stefan
Wirbelstromsensorbasierte Lokalisierung von Schienenfahrzeugen in topologischen Karten. 2011
ISBN 978-3-86644-749-3

Band 019   Carsten Hasberg
Simultane Lokalisierung und Kartierung spurgeführter Systeme. 2012
ISBN 978-3-86644-831-5

Band 020   Pitzer, Benjamin
Automatic Reconstruction of Textured 3D Models. 2012
ISBN 978-3-86644-805-6

Band 021   Roser, Martin
Modellbasierte und positionsgenaue Erkennung von Regentropfen in Bildfolgen zur Verbesserung von videobasierten Fahrerassistenzfunktionen. 2012
ISBN 978-3-86644-926-8

Band 022   Loose, Heidi
Dreidimensionale Straßenmodelle für Fahrerassistenzsysteme auf Landstraßen. 2013
ISBN 978-3-86644-942-8

Band 023   Rapp, Holger
Reconstruction of Specular Reflective Surfaces using Auto-Calibrating Deflectometry. 2013
ISBN 978-3-86644-966-4